Progress in Optical Science and Photonics

Volume 29

Indexed by Scopus

The purpose of the series Progress in Optical Science and Photonics is to provide a forum to disseminate the latest research findings in various areas of Optics and its applications. The intended audience are physicists, electrical and electronic engineers, applied mathematicians, biomedical engineers, and advanced graduate students.

Mengtao Sun · Xijiao Mu · Rui Li

Linear and Nonlinear Optical Spectroscopy and Microscopy

清华大学出版社
TSINGHUA UNIVERSITY PRESS

Springer

Mengtao Sun
University of Science and Technology
Beijing
Beijing, China

Xijiao Mu
University of Science and Technology
Beijing
Beijing, China

Rui Li
University of Science and Technology
Beijing
Beijing, China

ISSN 2363-5096 ISSN 2363-510X (electronic)
Progress in Optical Science and Photonics
ISBN 978-981-99-3636-6 ISBN 978-981-99-3637-3 (eBook)
https://doi.org/10.1007/978-981-99-3637-3

Jointly published with Tsinghua University Press
The print edition is not for sale in China (Mainland). Customers from China (Mainland) please order the print book from: Tsinghua University Press.

This work was supported by the National Natural Science Foundation of China (91436102, 11374353 and 11874084), and the fundamental Research Funds for the Central Universities.

This Springer imprint is published by the registered company Springer Nature Singapore Pte Ltd.
The registered company address is: 152 Beach Road, #21-01/04 Gateway East, Singapore 189721, Singapore

Paper in this product is recyclable.

Contents

Chapter 1
Introduction

Optical microscopy is unique in current imaging modalities. It can detect living tissue at subcellular resolution, visualize morphological details in tissue, and cannot be resolved by ultrasound or magnetic resonance imaging. However, to date, optical microscopy has not been fully successful in providing high resolution morphological information with chemical specificity. For example, the contrast of the confocal reflectance microscope [1] and the optical coherence spectrum [2] is based on the refractive index difference and cannot directly detect the chemical composition of the tissue structure. Fluorescence microscopy, although extremely sensitive and widely used, is limited in chemoselectivity due to the small number of intrinsic fluorophores such as NAD(P)H, riboflavin and elastin [3]. The introduction of an exogenous fluorophore provides a specific probe, but often causes undesirable perturbations. Second harmonic generation microscopy can be used to visualize well-ordered protein components, such as collagen fibers, but with insufficient sensitivity and specificity for other tissue components [4]. The vibrational spectrum of a biological sample contains multiple molecular features that can be used to identify biochemical components in the tissue. However, conventional vibration microscopy methods lack the sensitivity required for rapid tissue inspection. Infrared microscopy is limited by the low spatial resolution caused by long-wavelength infrared light [5] and the strong water absorption in biological samples. Although Raman microspectroscopy can distinguish between healthy and diseased tissues in vivo [5], it is hindered by undesired long integration times and/or high laser power in biomedical applications. A stronger vibration signal [6] can be obtained by coherent anti-Stokes Raman scattering (CARS), a nonlinear Raman technique. Typical CARS signals from submicron objects are orders of magnitude stronger than the corresponding spontaneous Raman responses. Since CARS is a non-linear effect, the signal is only generated at the laser focus, which allows point-by-point 3D imaging of thick samples, similar to a two-photon fluorescence microscope [7]. Recent developments in laser sources and detection protocols have significantly improved the ability of CARS as a bioimaging

M. Sun et al., *Linear and Nonlinear Optical Spectroscopy and Microscopy*,
Progress in Optical Science and Photonics 29,
https://doi.org/10.1007/978-981-99-3637-3_1

model [8]. CARS microscopy has been shown to be useful for mapping lipid compartments [9], protein clusters [10] and water distribution [11] in cell tissue cultures.

Precise imaging and treatment of biological tissues on a microscopic scale is a major requirement of modern biomedicine and clinical medicine. In recent years, optical imaging and diagnosis technology has achieved rapid development and considerable progress. For example, the confocal laser scanning microscope excites the fluorescent label on the two-dimensional plane in the tissue by means of point irradiation, which has become a widely used tool in biomedical imaging [12]; the super-resolution optical imaging method has broken through the optical diffraction limit and can obtain nano The scale of spatial resolution can observe the dynamic behavior of subcellular structures and biomolecules in living cells, which greatly promotes the development of cell biology [13, 14]. However, these technologies all use fluorescent dyes, fluorescent nanocrystals or genetically encoded fluorescent proteins. The properties of these fluorescent substances have inherent limitations on optical imaging: (1) Dye saturation. The maximum number of photons that a fluorescent dye can emit in a given time is limited. Only when an excited electron occupies the excited state for about 5 ns and then returns to the ground state, the fluorescence re-emission of the dye molecule will occur. (2) Dye bleaching. The total number of photons that fluorescent dyes can emit is limited. When the excited dye molecule changes from a singlet state to a triplet state, the chemical damage process of the dye will occur. The dye molecules in this state are highly reactive with oxygen molecules, which produces singlet oxygen (i.e. free radicals), which permanently destroy the dye and become the main source of phototoxicity affecting cells and tissues. (3) Fluorescence flicker. Most fluorescent dye molecules will "turn on" and "turn off" intermittently even under continuous light. The existence of the "off" period limits the long-term fluorescence imaging tracing process, causing the labeled molecules to be unable to be tracked continuously. Flicker and dye saturation greatly limit the number of photons that can be detected in a given period of time, resulting in a decrease in image signal-to-noise ratio. In addition, the phototoxicity and photobleaching of fluorescent materials also limit the length of time that biological targets can be observed [15].

With the increasing application of femtosecond laser technology, nonlinear optical imaging technology with femtosecond pulsed laser as the light source has aroused great interest of researchers. Non-linear imaging methods that have received widespread attention include: two-photon fluorescence (TPEF) imaging, coherent anti-Stokes Raman (CARS) imaging, and second harmonic (SHG) imaging [7, 16–18]. Two-photon and multi-photon fluorescence imaging requires the use of fluorescent probes, but the emission wavelength of high-quantum-efficiency fluorescent probes is mainly in the visible light region, and the fluorescent signal in the visible light region is still subject to strong tissue scattering and absorption [19]. In CARS imaging, the Raman signal of the molecular vibration spectrum is usually much lower than the fluorescence signal, [20, 21] which makes it more limited in detection sensitivity, acquisition time, laser power, etc., which greatly affects CARS in living tissues. Applications in imaging. The second harmonic uses excitation light in the near-infrared region (700 1300 nm), with a detection depth of up to 1000 μm, low phototoxicity and point excitation, so it is more suitable for living tissue and animal imaging. Second harmonic imaging is a unique nonlinear optical imaging technology. It is fundamentally different from two-photon fluorescence and coherent anti-Stoklaman in terms of imaging principles. The main features are as follows: (1) SHG is a second-order nonlinear effect. TPEF and CARS are third-order nonlinear effects. In non-centrosymmetric materials, the second-order nonlinear effects are much larger than the third-order nonlinear effects. (2) SHG is a nonlinear scattering process, the sample does not absorb energy, which fundamentally overcomes phototoxicity and photodamage; while in two-photon imaging, fluorescent molecules suffer from severe photobleaching, and energy is lost due to processes such as vibration relaxation. (3) SHG signal is strictly frequency-doubled. Changing the pump wavelength can obtain frequency-doubled signals of different wavelengths. TPEF has a red shift relative to the double frequency, and the emission spectrum remains unchanged. Although the imaging principle is different, the second harmonic is fully compatible with the two-photon imaging system. Therefore, the two-photon confocal system can be easily transformed into a second harmonic imaging system by replacing the filter.

Hematoxylin and eosin (HE) staining is a standard histopathological method used as a clinical diagnosis [22]. Eosin stains proteins and cytoplasm in bright pink, while hematoxylin stains basic structures (such as DNA) in blue-violet. However, HE staining is a very slow process that requires biopsy, fixation, sectioning and staining. It usually takes several days, so it cannot be used for intraoperative diagnosis. Intraoperative freezing technology still takes a long time, at least 30 minutes. Therefore, fast and accurate imaging methods are very necessary for intraoperative diagnosis, and it is also very important for the judgment of the resection margin and the making of surgical decisions.

Research on detecting disease states through a series of technologies has been rapidly developed. Non-invasive imaging methods include computed tomography (CT), magnetic resonance imaging (MRI), and positron emission tomography (PET), [23] but they are largely affected by low spatial resolution and intraoperative Compatibility limitations. Intraoperative MRI can provide updated images during the operation, indicating that this technique has great potential, but it is limited due to the high cost and extended operation time [24]. Ultrasound (US) and optical coherence tomography (OCT) have been proven to provide structural information in real time, but they can only be used on a large scale, and it is difficult to obtain high-definition tissue structure information with subcellular resolution [25], and lack of molecular specificity [26].

In recent years, fluorescent imaging using molecular markers has made important breakthroughs and improved the sensitivity of intraoperative detection. For example, brain tumor imaging is expected to reveal the edge of brain tumors, but it is still subject to certain restrictions [27–29]: First, Only some cancer cells absorb fluorescent molecules [29]; secondly, the dye has the disadvantage of non-specific labeling; furthermore, fluorescein often undergoes fluorescent bleaching under laser irradiation. Confocal microscopy has been used for intraoperative imaging of fluorescently labeled tissues, but it also has limitations similar to fluorescent imaging [27]. Various nonlinear optical imaging techniques are also used for tissue imaging, such as second harmonic (SHG) and third harmonic microscopy (THG). Among them: second harmonic microscopy can selectively image structures with non-central inversion symmetry (such as collagen fibers and microtubules) [30]; third harmonic microscopy is sensitive to the non-uniformity of refractive index , But it cannot provide enough molecular information [31].

Vibration spectrum imaging provides a new method for specific imaging of pathological tissues. The fingerprint vibration spectra of molecules can be recorded by infrared spectroscopy or Raman spectroscopy [32–34]. However, spontaneous Raman imaging of biological tissues is limited by weak signal strength and slow imaging speed, and it is difficult to directly use in biomedical imaging. Although the surface-enhanced Raman scattering (SERS) method has the advantage of significantly enhancing the Raman signal, it requires the use of nanoparticles for exogenous labeling [35]. Stimulated Raman Scattering (SRS) microscopic imaging technology has been widely used in the field of biomedicine due to its unique chemical bond-specific imaging function, including label-free DNA imaging [36], drug molecule tracking [37, 38], Tumor detection [39, 40], lipid quantitative analysis [41, 42], molecular metabolism and the mechanism of action of biological enzymes, etc. [43, 44]. Compared with fluorescence imaging, a major advantage of stimulated Raman scattering imaging technology is that it does not require any markers to help it complete the imaging, and it has the advantages of high sensitivity, molecular selectivity and high resolution. Stimulated Raman Scattering Microscopy imaging technology overcomes the potential problems caused by label imaging technology with its characteristics of label-free imaging, such as non-specific labeling, toxicity, and influence on the biological process of the label [45].

Optical microscope has become an important tool for biomedical research by virtue of its high-resolution, non-contact, non-invasive, and fast imaging advantages. Every advancement in optical microscopy imaging technology has greatly promoted the development of life sciences, basic medicine and clinical diagnostics. Since the 20th century, the field of optical microscopy imaging has achieved rapid development, and many new technologies and methods have emerged, including confocal microscopy, two-photon microscopy, light sheet illumination microscopy, and super-resolution microscopy, and so on. Among these microscopic imaging technologies, two-photon microscopy imaging is one of the most milestone technologies, and fluorescence lifetime detection technology has opened up a new detection function for two-photon microscopy imaging.

Two-photon microscopy imaging technology was first implemented by Professor Webb [7] in 1990. This technology uses two-photon excitation fluorescence signals for three-dimensional microscopy imaging. The use of low-scattering near-infrared light for local excitation makes this technology has the advantages of low photobleaching and phototoxicity, super strong tissue penetration, subcellular level resolution, and inherent tomographic capabilities. In addition, this technology can also use endogenous optical markers to obtain contrast and achieve label-free imaging [3, 46, 47]. In view of the above advantages, two-photon microscopy imaging technology is considered to be one of the most suitable technologies for in vivo optical microscopy imaging [47]. This technology has gradually become a research method for the occurrence, development and potential treatment of diseases such as tumors and Alzheimer's disease. In addition, the two-photon microscopy imaging technology is non-invasive and can achieve label-free imaging characteristics, and its advantages in imaging depth and resolution make it one of the most promising clinical research tools. At present, this technology has been successfully used in clinical research on tumors, tissue lesions, controlled drug release, and in vivo drug screening [48].

Fluorescence lifetime detection refers to the use of time-resolved technology to detect the dynamic process of fluorescence intensity attenuation. Under the excitation of high-energy light, the fluorescent substance will transition to an unstable excited state, and radiate fluorescent photons when it returns to a stable ground state. Therefore, the fluorescence lifetime reflects the average time that the fluorescent substance stays in the excited state [49]. Similar to fluorescence spectroscopy, fluorescence lifetime is another important characteristic of fluorescent materials. Fluorescence lifetime detection breaks through the limitations of traditional steady-state fluorescence detection and adds an independent dimension of new information to fluorescence imaging [50–52]. Since the transition process of the fluorophore from the excited state to the ground state is very easily affected by the local environment of the molecule, the fluorescence lifetime can also sensitively reflect the pH, temperature, oxygen concentration, ion concentration, enzyme activity, and molecular configuration of the environment in which the molecule is located [50–53]. For example, the free-state reduced nicotinamide adenine dinucleotide (NADH) has a fluorescence lifetime of several hundred picoseconds, while protein-bound NADH has a fluorescence lifetime of several nanoseconds [54]. In addition, the fluorescence lifetime measurement is usually independent of the concentration and quantum yield of fluorescent substances [52, 53, 55]. Therefore, the use of time-resolved fluorescence detection technology to study biological systems has many unique advantages: (1) The fluorescence lifetime characteristics provide an additional contrast parameter for distinguishing fluorescent materials with overlapping emission spectra, so that biomolecules with overlapping spectra but different fluorescence decay times can be obtained. Effective resolution [51]; (2) The sensitivity of fluorescence lifetime measurement to various parameters of the microenvironment of biological tissues enables it to be used to detect local environmental parameters and study the interaction between proteins [51, 53]; (3) The concentration of fluorescent substances in tissues is usually unknown and constantly changing, so the fluorescence lifetime is independent of fluorescent substance concentration and quantum yield characteristics, which makes it possible to achieve more accurate *in vivo* quantitative measurements compared to steady-state fluorescence detection technology [51, 53]. Based on the above advantages, fluorescence lifetime imaging technology has received great attention from researchers, and this technology has been widely used in biophysics and medical diagnostic research [56].

The combination of two-photon microscopy imaging technology and fluorescence lifetime detection technology creates a win-win situation with complementary advantages. On the one hand, the two-photon microscopy imaging technology can not only provide the pulsed excitation light source required for fluorescence lifetime detection, but also benefit from its inherent tomographic ability. This technology can effectively avoid signals of different depths when performing fluorescence lifetime measurement on thick tissues. On the other hand, the combination with the fluorescence lifetime detection technology enables the two-photon microscopy imaging technology to provide multiple-dimensional fluorescence signal detection modes, and both the function and the application range have been expanded. Biomedicine and clinical diagnosis provide a new research method [52]. Many leading scientific research teams in the world have been actively carrying out research on instruments and equipment, data analysis techniques, and biomedical and clinical applications related to two-photon fluorescence lifetime imaging, and have made many breakthroughs. The country is still in its infancy, and only a few scientific research teams are engaged in related research. Among them, the team of Professor Qu Junle of Shenzhen University has long been committed to the research of two-photon fluorescence lifetime imaging technology and its application in biomedicine. At present, the technology has been applied to the research of tumor mechanism and diagnostic methods and the research of molecular diagnostic technology [57]. Our research group also has deep accumulation in this area, and is currently conducting research on the diagnosis of benign and malignant diseases of the digestive tract and brain tumors based on two-photon fluorescence lifetime. This article will briefly summarize the concept of two-photon fluorescence lifetime and common detection methods, combined with the latest research results of this research group, summarize the research progress of two-photon fluorescence lifetime imaging in tumor detection, and finally look forward to the future clinical application of this technology. Challenges and potential advantages.

The nonlinear optical spectrum signal is a new type of optical characterization technology due to the non-invasiveness and good biocompatibility of the measurement technology. Since the Stokes Raman is often affected by the fluorescence effect of the detection object, in order to avoid the fluorescent signal, based on the characteristic that the frequency of the Stokes Raman signal is higher than the frequency of the fluorescent signal, the anti-Stokes Raman measurement The method was proposed, but the anti-Stokes signal was far smaller than the Stokes signal. As a result, Coherent Anti-Stokes Raman Spectroscopy (CARS) and imaging technology came into being under the unremitting efforts of scientists, and successfully achieved signal measurement. A technique similar to CARS is stimulated Raman scattering (SRS). At the same time, there are two other important nonlinear optical signals and their microscopic imaging methods. They are two-photon excited fluorescence (TPEF) and second harmonic generation (SHG). The principles and experiments of these four types of nonlinear optical signals, especially two-dimensional and three-dimensional imaging, have great application potential in materials, chemistry and biomedicine. This book focuses on introduction to the principles of SRS, CARS, TPEF and SHG signals, and appropriate theoretical calculation methods. As well as these signal measurement, imaging specific methods and experimental results and their analysis. The theoretical part starts from the basic theory of nonlinear optics and the relationship between strong light, and gradually transitions to specific theoretical calculation methods for specific optical signals. The theoretical part contains the combination of classical theory and quantum theory, so that readers can understand the core of these technologies well. The experimental part focuses on the introduction of the experimental technology of spectroscopy and the experimental part of imaging, mainly to enable readers to have a more comprehensive understanding of nonlinear optical signal spectra and imaging methods. The experimental part mainly introduces the application of nonlinear optical spectroscopy and imaging technology in the fields of materials and biology. This book combines recent high-quality scientific research results in the field of nonlinear optics at home and abroad as well as my years of research results and experience in this field. This book focuses on both theory and experiment of nonlinear optical signals and imaging. The theoretical part is not only an introduction to the basic theory, but also as much as possible to introduce practical calculation methods, so that the theoretical part can also be applied. The experimental part mainly focuses on the introduction of imaging technology and its applications in materials, chemistry and biology. This book focuses on both nonlinear optical technology and theory. First of all, we will first elaborate on the physical principles underlying this technology before the description of each nonlinear optical technology. The principle is roughly divided into two parts, and the theory is explained and discussed in depth from the perspectives of classical theory and quantum theory. In the theoretical part, we also introduced the methods of these theories in actual calculation and analysis and the corresponding calculation theories. Therefore, the theoretical part can be used as a guide for the theoretical calculation of nonlinear optical technology. Secondly, we also review the experimental methods and excellent results of nonlinear optical spectroscopy and imaging technology in the world. Introduce well-known scientists and the latest experimental and theoretical results obtained by our group in the field of nonlinear optics and spectral imaging, and add years of experience and knowledge in scientific research in related fields. In the end, the whole book has a unique way of explaining the theory and experimental methods of nonlinear optical spectroscopy and imaging in a small but precise manner, rather than a broad introduction to all nonlinear optics.

Chapter 2
Basic Theory of Nonlinear Optics

A branch of modern optics that studies the nonlinear phenomena produced by strong media in the presence of strong coherent light and its applications. The study of nonlinear optics is of great significance to laser technology, spectroscopy development, and material structure analysis. The field of nonlinear optics is mainly concerned with various new optical phenomena and effects that occur during the interaction between intense laser radiation and matter, including in-depth understanding and exploration of these new phenomena and the causes of the new process [58]. Before the advent of lasers, some important formulas describing common optical phenomena were often linear. The polarization strength vector of the medium is such a very important physical quantity, which describes the important phenomena such as dispersion and scattering of light during the propagation of the medium. Before the appearance of a strong field laser, it is assumed to have a simple linear relationship with the incident electric field strength $\overrightarrow{E}$:

$$P = \epsilon_0 \chi \overrightarrow{E} \tag{2.0.1}$$

where the ϵ_0 is the vacuum dielectric constant; the coefficient χ is the dielectric polarization of the medium. From this perspective, in the theoretical framework of classical electrodynamics, the macroscopic Maxwell's equations describing the interaction between light and matter are also a set of linear partial differential equations. In other words, there is only a linear term in the equation that contains the field strength vector. Therefore, deductive reasoning can be seen that a bundle of monochromatic light is incident on the medium, and its frequency does not change; when light of different frequencies is incident at the same time, mutual coupling does not occur, and no new light is generated. But the above conclusions were fundamentally shaken in 1960. This year the world's first laser, the ruby laser, was born. The scientists used a 694.3 nm laser output from a pulsed ruby laser to enter the quartz crystal. For the first time, 347.2 nm multiplier coherent radiation was observed. After this incident, different abnormal optical phenomena have sprung up. In a short period of time, people have observed second harmonics, third harmonics, and optical and frequency in a

M. Sun et al., *Linear and Nonlinear Optical Spectroscopy and Microscopy*,
Progress in Optical Science and Photonics 29,
https://doi.org/10.1007/978-981-99-3637-3_2

series of media. Scientists pointed out that as long as the previous linear polarization theory is extended to higher order, the new effect can be perfectly explained. At this point, the polarization of the dielectric is no longer a simple linear relationship with the incident light field, but a higher order power series relationship, namely:

$$P = \epsilon_0 \left[\chi^{(1)} \vec{E} + \chi^{(2)} \vec{E}\vec{E} + \chi^{(3)} \vec{E}\vec{E}\vec{E} + \cdots \right] \tag{2.0.2}$$

where $\chi^{(1)}$, $\chi^{(2)}$ and $\chi^{(3)}$ are the first-order (linear), second-order (non-linear) and third-order (non-linear) polarization rates of the medium, respectively. In general, they all appear as tensor forms. By introducing the above representation of the polarization strength into Maxwell's equations, a set of nonlinear electromagnetic wave equations containing higher field strength terms can be derived. It is thus possible to explain the generation of frequency doubling radiation when a single frequency light is incident on a particular medium.

2.1 Classical Electromagnetic Theory of Nonlinear Optics

2.1.1 Measurement of Nonlinear Optical Processes

If the relationship (2.0.1) is established in an isotropic medium or an anisotropic medium, when a monochromatic light wave is incident on the medium, the polarization intensity P is harmonically changed at the same frequency as E. And radiate electromagnetic waves of the same frequency, which is the secondary wave radiation. The results obtained by the interaction of the secondary wave radiation with the incident light wave can explain the reflection, refraction and scattering of light. The linear wave equation in a homogeneous isotropic medium is:

$$\Delta^2 \vec{E} - \mu\epsilon \frac{\partial^2 \vec{E}}{\partial t^2} = 0 \tag{2.1.1}$$

The above equation is a set of linear differential equations of three components of $\vec{E}$, so that when light of different frequencies is simultaneously incident into the medium, no mutual coupling occurs between the light waves, and thus no new frequency of light is generated. In an anisotropic medium, the dielectric constant in (2.1.1) should be the dielectric tensor ϵ, at which time there is a cross-interaction between the three components of the electric field of a certain frequency of the vector E that constitutes vector $\vec{E}$. However, since there is only one E of the wave equation, the different frequency components still function independently. There is no cross-coupling between the waves of different frequencies, and they each satisfy a wave equation.

2.1.2 Nonlinear Induced Polarization Effect of Optical Media

Many typical and important nonlinear optical effects, such as optical frequency doubling, and frequency and difference frequency, optical parametric amplification and autofocus, can be used in the framework of semi-classical theory, using optical media under the action of strong light. The theory of the linear polarization process is explained and processed.

For a non-resonant-absorbing transparent optical medium, the atoms, molecules or ions that make up the medium do not undergo transitions between their different quantum mechanical intrinsic energy levels under the action of an optical monochromatic electromagnetic field. However, the distribution and motion state of the internal charge of these particles will change with a certain condition relative to the case of no external electromagnetic field perturbation, which will cause the electric dipole moment of the light field induction, and the latter constitute a new electromagnetic wave radiation source. When describing this particular source of radiation, it is necessary to introduce the induced dipole moment of the atom. The polarization strength vector $\boldsymbol{P}$ of the medium is defined as the sum of the induced electric dipole moment vectors within the unit volume of the medium. Let the number of atoms in the unit volume of the medium be N, and the induced electric dipole moment vector of the i-th atom is $\boldsymbol{p}_i$, then the overall induced electric dipole moment is expressed as:

$$\boldsymbol{P}(t) = \sum_{i=1}^{N} \boldsymbol{p}_i(t) \tag{2.1.2}$$

Since the action square is an optical frequency electromagnetic wave, in general, $\boldsymbol{p}_i$ and $\boldsymbol{P}$ are functions as a function of time. It can be seen from (2.1.2) that the polarization strength of the medium is determined by two factors: first, the induced electric dipole moment of a single atom constituting the medium under the disturbance of the light field, and secondly the induced electric dipole moment vector between the superposition of different atoms. In the case where the ambient incident light field condition is given, the induced electric dipole moment of a single atom or molecule in the medium is mainly determined by the molecular structure, that is, the wave function of quantum mechanics. In summary, the induced electric dipole moment of the medium is a function of time and external field disturbances. Therefore, we can illuminate the polarization intensity vector $\boldsymbol{P}$ phenotypically as follows:

$$P(t) = P^{(1)}(t) + P^{(2)}(t) + P^{(3)}(t) + \cdots + P^{(i)}(t) + \cdots \tag{2.1.3}$$

Assuming that the nth term in the series expansion is only related to the nth power function of the field strength $\boldsymbol{E}(t)$, which is also an arbitrary function of time, it can be written as the following multiple integral form in general.

$$\boldsymbol{P}^{(n)}(t) = \varepsilon_0 \int_{-\infty}^{\infty} \mathrm{d}t_1 \int_{-\infty}^{\infty} \mathrm{d}t_2 \cdots \int_{-\infty}^{\infty} \mathrm{d}t_n \cdot \boldsymbol{R}^{(n)}(t; t_1, t_2, \ldots, t_n) \\ \boldsymbol{E}(t_1)\, \boldsymbol{E}(t_1)\, \boldsymbol{E}(t_2) \cdots \boldsymbol{E}(t_n) \tag{2.1.4}$$

where $R^{(n)}(t; t_1, t_2, \ldots, t_n)$ is the nth-order polarization response function of the medium to the light field. And it is an n+1 order tensor. In general, the electrode's electrode response is independent of the choice of time, so the above equation can be simplified as:

$$\boldsymbol{P}^{(n)}(t) = \varepsilon_0 \int_{-\infty}^{\infty} \mathrm{d}t_1 \int_{-\infty}^{\infty} \mathrm{d}t_2 \cdots \int_{-\infty}^{\infty} \mathrm{d}t_n \cdot \boldsymbol{R}^{(n)}(t_1, t_2, \ldots, t_n)\, \boldsymbol{E}(t - t_1) \\ \boldsymbol{E}(t - t_2) \cdots E(t - t_n) \tag{2.1.5}$$

According to the principle of Fourier analysis, any time function can be expressed as a form of Fourier series or Fourier integral of a particular form. So in general we can express the electric field and the polarization strength as the form of Fourier integral:

$$\left.\begin{array}{l} \boldsymbol{E}(t) = \int_{-\infty}^{\infty} \mathrm{d}\omega \boldsymbol{E}(\omega) \mathrm{e}^{-\mathrm{i}\omega t} \\ \boldsymbol{P}^{(n)}(t) = \int_{-\infty}^{\infty} \mathrm{d}\omega \boldsymbol{P}^{(n)}(\omega) \mathrm{e}^{-\mathrm{i}\omega t} \end{array}\right\} \tag{2.1.6}$$

By relating this relationship to the above response function, a general relationship between the polarization vector $\boldsymbol{P}(t)$ and the electric field $\boldsymbol{E}(t)$ can be obtained:

$$\boldsymbol{P}^{(n)}(t) = \varepsilon_0 \int_{-\infty}^{\infty} \mathrm{d}\omega_1 \cdots \int_{-\infty}^{\infty} \mathrm{d}\omega_n \cdot \chi^{(n)}(\omega_1, \ldots, \omega_n)\, E(\omega_1) \cdots E(\omega_n) \cdot \exp\left[-\mathrm{i}t \sum_{m=1}^{n} \omega_m\right] \tag{2.1.7}$$

where $\chi^{(n)}(\omega_1, \ldots, \omega_n)$ is the nth-order polarization of the optical medium, which is the form of the n+1 order tensor. This tensor is related to the electrode response function $\boldsymbol{R}^{(n)}(t; t_1, t_2, \ldots, t_n)$:

$$\chi^{(n)}(\omega_1, \ldots, \omega_n) = \int_{-\infty}^{\infty} \mathrm{d}t_1 \cdots \int_{-\infty}^{\infty} \mathrm{d}t_n \cdot \boldsymbol{R}^{(n)}(t_1, \ldots, t_n) \cdot \exp\left[\mathrm{i} \sum_{m=1}^{n} \omega_m t_m\right] \tag{2.1.8}$$

2.1.3 Tensor Representation of Nonlinear Polarization

From an experimental point of view, most of the nonlinear optical effects are excited by monochromatic light or quasi-monochromatic pulsed lasers. In this case, the incident field can be thought of as consisting of a single frequency monochromatic Fourier component, and the polarization intensity is generated by the mutual coupling of the incident light field in the medium. This means that we only need to use the

$\chi^{(n)}$ dielectric ordering rate of the medium to describe the above-mentioned coupling process is sufficient. As mentioned above, the nth-order polarization rate of the medium is in the form of n+1 order tensor. The form and operation rules of these tensors are described below. In the first approximation, the Fourier component of the linear polarization of the medium is expressed as:

$$\boldsymbol{P}^{(1)}(\omega) = \varepsilon_0 \chi^{(1)}(\omega) \boldsymbol{E}(\omega) \tag{2.1.9}$$

This relationship represents a simple linear relationship between the polarization strength and the incident field. An electromagnetic field of frequency ω can only cause radiation of the same frequency when incident. From this relationship, the characteristics of reflection, refraction, scattering, and refractive index dispersion of ordinary light can be explained. According to the foregoing, for a general anisotropic medium, $\chi^{(1)}$ generally takes a second order tensor with 3*3 tensor elements. Using the matrix of tensors to represent the above formula can be rewritten as:

$$\begin{pmatrix} P_x^{(1)}(\omega) \\ P_y^{(1)}(\omega) \\ P_z^{(1)}(\omega) \end{pmatrix} = \varepsilon_0 \begin{pmatrix} \chi_{xx}^{(1)}(\omega) & \chi_{xy}^{(1)}(\omega) & \chi_{xz}^{(1)}(\omega) \\ \chi_{yx}^{(1)}(\omega) & \chi_{yy}^{(1)}(\omega) & \chi_{yz}^{(1)}(\omega) \\ \chi_{zx}^{(1)}(\omega) & \chi_{zy}^{(1)}(\omega) & \chi_{zz}^{(1)}(\omega) \end{pmatrix} \begin{pmatrix} E_x(\omega) \\ E_y(\omega) \\ E_z(\omega) \end{pmatrix} \tag{2.1.10}$$

In the formula, $P_x^{(1)}(\omega)$, $E_x(\omega)$, etc. are the projection components of the two vectors $P^{(1)}(\omega)$ and $E(\omega)$ in a Cartesian coordinate system. Therefore, (2.1.10) can also be expressed equivalently as a summation formula:

$$P_i^{(1)}(\omega) = \varepsilon_0 \sum_j \chi_{ij}^{(1)}(\omega) E_j(\omega), \quad i, j = x, y, z \tag{2.1.11}$$

According to (2.1.9), the second-order nonlinear polarization intensity Fourier component can be written as:

$$\boldsymbol{P}^{(2)}(\omega = \omega_1 + \omega_2) = \varepsilon_0 \chi^{(2)}(\omega_1, \omega_2) E(\omega_1) E(\omega_2) \tag{2.1.12}$$

The physical meaning of the above formula is two monochromatic light fields of frequency ω_1 and ω_2 under the second approximation. Through the second-order nonlinear process, the coherent radiation at the new frequency $\omega = \omega_1 + \omega_2$ can be induced in the medium. The second-order nonlinear polarization rate $\chi^{(2)}$ is a third-order tensor, and there are twenty-seven tensor elements. Correspondingly, the matrix representation of the tensor can be written as:

$$\begin{pmatrix} P_x^{(2)}(\omega) \\ P_y^{(2)}(\omega) \\ P_z^{(2)}(\omega) \end{pmatrix} = \varepsilon_0 \begin{pmatrix} \chi_{xxx}^{(2)} \; \chi_{xxy}^{(2)} \cdots \chi_{xzz}^{(2)} \\ \chi_{yxx}^{(2)} \; \chi_{yxy}^{(2)} \cdots \chi_{yzz}^{(2)} \\ \chi_{zxx}^{(2)} \; \chi_{zxy}^{(2)} \cdots \chi_{zzz}^{(2)} \end{pmatrix} \begin{pmatrix} E_x(\omega_1) E_x(\omega_2) \\ E_x(\omega_1) E_y(\omega_2) \\ E_x(\omega_1) E_z(\omega_2) \\ E_y(\omega_1) E_x(\omega_2) \\ E_y(\omega_1) E_y(\omega_2) \\ E_y(\omega_1) E_z(\omega_2) \\ E_z(\omega_1) E_x(\omega_2) \\ E_z(\omega_1) E_y(\omega_2) \\ E_z(\omega_1) E_z(\omega_2) \end{pmatrix} \tag{2.1.13}$$

Of course, this tensor representation can be expressed as a summation convention:

$$P_i^{(2)}(\omega = \omega_1 + \omega_2) = \varepsilon_0 \sum_{jk} \chi_{ijk}^{(2)}(\omega_1, \omega_2) E_j(\omega_1) E_k(\omega_2), \quad i, j, k = x, y, z \tag{2.1.14}$$

The above is the tensor representation of linear and nonlinear polarizabilities. This matrix is very important for quantum mechanical calculations to analyze nonlinear processes.

2.1.4 Rotational Symmetry of Nonlinear Polarizability Tensor Elements

In the previous section, we introduced the specific representation of the tensor of vitrification. In this section, we will introduce the rotational symmetry of the polarizability tensor element in combination with the intrinsic properties of the tensor. As the order of the tensor increases, the tensor element becomes very much. In order to reduce the complexity of analysis and calculation, it is necessary to introduce rotational symmetry. For the second and third order polarizability tensor elements, the symmetry of the following tensor indicators are respectively:

$$\begin{aligned} &\chi_{ijk}^{(2)}(\omega_1, \omega_2) = \chi_{ikj}^{(2)}(\omega_2, \omega_1) \\ &\chi_{ijkl}^{(3)}(\omega_1, \omega_2, \omega_3) = \chi_{ikjl}^{(3)}(\omega_2, \omega_1, \omega_3) = \chi_{iljk}^{(3)}(\omega_3, \omega_1, \omega_2) = \cdots \end{aligned} \tag{2.1.15}$$

The meaning in the above formula is that the nonlinear polarizability tensor element is arbitrarily interchanged with the corresponding frequency position $(\omega_1, \omega_2, \omega_3, \ldots)$ for other indicators $(j, k, \ldots)$ except the first index i, and the tensor element value is unchanged.Under the premise of the above symmetry, a frequency hidden variable $\omega' = -(\omega_1 + \omega_2 + \omega_3 + \cdots)$ is introduced, then the range of rotation symmetry is expanded:

$$\chi_{ijk}^{(2)}\left[\omega' = -(\omega_1+\omega_2);\omega_1,\omega_2\right] = \chi_{jik}^{(2)}\left(\omega_1;\omega',\omega_2\right) = \chi_{kji}^{(2)}\left(\omega_2;\omega_1,\omega'\right)$$
$$\chi_{ijkl}^{(3)}\left[\omega' = -(\omega_1+\omega_2+\omega_3);\omega_1,\omega_2,\omega_3\right] = \chi_{jikl}^{(3)}\left(\omega_1;\omega',\omega_2,\omega_3\right) = \cdots \quad (2.1.16)$$

The meaning in the above formula is that the nonlinear polarizability tensor element is arbitrarily interchanged with the corresponding frequency position ($\omega', \omega_1, \omega_2, \omega_3, \ldots$) for the indicators ($i, j, k, \ldots$), and the tensor element value is unchanged.

2.1.5 *Time Reversal Symmetry of Polarization Rate*

The nonlinear polarizability of the medium is generally expressed in the form of a complex number, that is, it can be expressed in the form of a real part plus an imaginary part. Each part has a different physical meaning, and the imaginary part generally indicates the absorption of the electromagnetic field by the medium. For example, the imaginary part of the third-order nonlinear polarizability represents the two-photon absorption coefficient of the medium. Therefore, according to the nature of the polarization response function, the complex conjugate operation of $\chi^{(n)}$ has the following relationship:

$$\left[\chi^{(n)}(\omega_1,\omega_2,\ldots,\omega_n)\right]^* = \chi^{(n)}(-\omega_1,-\omega_2,\ldots,-\omega_n) \quad (2.1.17)$$

It can be further proved that under the premise of non-resonance, the approximate polarization of the optical medium can be considered as a real number:

$$\left[\chi^{(n)}(\omega_1,\omega_2,\ldots,\omega_n)\right]^* = \chi^{(n)}(\omega_1,\omega_2,\ldots,\omega_n) \quad (2.1.18)$$

Combining (2.1.17) and (2.1.18) gives the following relationship:

$$\chi^{(n)}(\omega_1,\omega_2,\ldots,\omega_n) = \chi^{(n)}(-\omega_1,-\omega_2,\ldots,-\omega_n) \quad (2.1.19)$$

The physical meaning of the above relationship is that the nonlinear polarizability remains constant relative to the overall inverse of all its frequencies. This is the time reversal symmetry of the polarizability. These symmetries can reduce the number of tensor elements necessary for a particular nonlinear optical process, which greatly simplifies the complexity of analysis and calculation.

2.2 Quantum Theory and Method of Nonlinear Optics

In the previous section we introduced the main forms of nonlinear polarizability and tensor representations based on classical physics and Fourier analysis. However, the tensor element of the nonlinear polarizability is related to the microstructure of the medium. At the microscopic scale, the classical physics paradigm does not apply. Therefore, quantum mechanics is needed to describe the nonlinear polarizability in the microscopic scale. The nonlinear polarizability is the response of the medium to the external field disturbance, so some response theories are applied in the field of calculating the nonlinear polarizability. On the other hand, the interaction of the medium and the external electromagnetic field is essentially the interaction of light with matter. Essentially, this is a quantum electrodynamic (QED) process. It describes the interaction of the light field (the bose field) with the electrons (fermions) in the material. This is a branch of QED, also known as quantum optics. In quantum optics, the density matrix and its equation of motion are generally used to describe the interaction of light with matter. This is due to the statistical physics method derived from the difficulty of solving the exact wave function of a multibody electronic system.

2.2.1 Density Matrix

When calculating the average of operators without knowing the exact wave function of the system, it is often necessary to use the density matrix method. First, let's examine the average of any Hermitian operator:

$$
\begin{aligned}
<\hat{A}>&=\int \psi^*(r,t)\hat{A}\psi(r,t)dv \\
&=\sum_{m,n} c_m^*(t)\left(\int u_m^*(r)\hat{A}u_n(r)dv\right)c_n(t) \\
&=\sum_{m,n} c_m^* c_n A_{mn}
\end{aligned}
\tag{2.2.1}
$$

where $c_n(t)$ is the time-dependent coefficient in the expansion of the wave function $\psi(r,t)=\sum c_n(t)u_n(r)$; A_{mn} is the matrix element of the operator: $A_{mn}=\int u_m^*(r)\hat{A}u_n(r)d\tau$. Assuming that the exact state $c_n(t)$ of the system is unknown, there is enough information to calculate the $c_m^* c_n$ ensemble average, and the ensemble average of the operator A average can be obtained:

$$
\overline{<A>}=\sum_{m,n}\overline{c_m^* c_n}A_{mn}
\tag{2.2.2}
$$

where $\rho_{mm}=\overline{c_m^* c_n}$ is defined as the density matrix. So the above formula can be written as the ensemble average of the mean of the operator A.

$$
\begin{aligned}
\overline{<A}&=\sum_{m,n} c_m^* c_n A_{mn}=\sum_n \rho_{mm}A_{mn}=\sum_n(\rho A)_m \\
&=\mathrm{tr}(\rho A)
\end{aligned}
\tag{2.2.3}
$$

The $\overline{c_m^* c_n}$ is the average of N systems of ensemble.

$$\rho_{nm} = \overline{c_m^* c_n} = \frac{1}{N} \sum_{s=1}^{N} \left(c_m^s\right)^* c_n^s \tag{2.2.4}$$

The density matrix is used to describe the probability characteristics of the ensemble state. The diagonal term ρ_{mm} describes the probability that a system in the ensemble is in an u_n. The non-diagonal element represents the ensemble average of $c_m^* c_n$. When discussing the interaction between light and matter. The light field induced atomic polarization and density matrix are only related to the density matrix, without the need to know the exact system wave function.

2.2.2 Time-Dependent Density Matrix

For the eigenfunction $\psi(r, t) = \sum c_n(t) u_n(r)$, the eigenequation with the inclusion time can be obtained:

$$i\hbar \frac{\partial}{\partial t} c_m(t) = \sum_n c_n(t) H_{mn} \tag{2.2.5}$$

At the same time, the definition of the density matrix $\rho_{mm} = \overline{c_m^* c_n}$ is the derivative of time:

$$\frac{\partial \rho_{nm}}{\partial t} = c_n \frac{\partial c_m^*}{\partial t} + c_m^* \frac{\partial c_n}{\partial t} \tag{2.2.6}$$

Combining the eigenequation and the total differential of the density matrix can be derived:

$$\begin{aligned}
\frac{\partial \rho_{nm}}{\partial t} &= \frac{i}{\hbar} \overline{c_n \sum_k c_k^* H_{mk}^+} - \frac{i}{\hbar} \overline{c_m^* \sum_k c_k H_{nk}} \\
&= \frac{i}{\hbar} \left(\overline{\sum_k c_n c_k^* H_{mk}^+} - \overline{\sum_k c_m^* c_k H_{nk}} \right) \\
&= \frac{i}{\hbar} \sum_k \left(\rho_{nk} H_{km} - H_{nk} \rho_{km} \right) \\
&= \frac{i}{\hbar} \left[(\rho H)_{mm} - (H\rho)_{nm} \right]
\end{aligned} \tag{2.2.7}$$

The above conclusion can be written in a very simple form with quantum Poisson brackets:

$$\frac{\partial \rho}{\partial t} = \frac{i}{\hbar} [\rho, H] \tag{2.2.8}$$

The quantum Poisson brackets are the Lie brackets. Meet the nature of the Lie brackets. Using the relationship of (2.2.8), the evolution of the density matrix in the process of density matrix and Hamiltonian interaction can be solved and analyzed. This is very important for nonlinear optics. Since the light intensity is weak (the light field is in an uncompressed state), this interaction is not strong, so the density matrix changes little. So at the macro level it seems that this process is linear. However, if the light intensity is strong, the light field is in a compact state, and the interaction between the density matrix and the Hamiltonian is strong. At this point, the density matrix changes greatly. Macroscopically, it exhibits nonlinear properties. Of course, in nonlinear optical experiments, short-pulse lasers are generally used, so the evolution of the density matrix over time is also very important for interpreting nonlinear optical experiments.

2.2.3 The Tensor and Properties of the Polarizability of the Independent Molecular System

Suppose there are M molecules in the volume V, where the unperturbed Hamiltonian and electric dipole moment operators of the mth molecule are $\hat{H}_m$ and $\hat{R}_m$, respectively, then the undisturbed Hamiltonian and galvanic couple of the entire set. The polar moment operator is:

$$\begin{aligned} \hat{H}_0 &= \sum\nolimits_m \hat{H}_m \\ \hat{R} &= \sum\nolimits_m \hat{R}_m \end{aligned} \tag{2.2.9}$$

Because the Hamiltonian of a single molecule is commutability, the density operator of the entire set in the thermal equilibrium state is:

$$\hat{\rho}_0 = Ae^{-\hat{H}_0/KT} = \hat{\rho}_1\hat{\rho}_2\cdots\hat{\rho}_m\cdots\hat{\rho}_M \tag{2.2.10}$$

where, $\hat{\rho}_m = A^{1/M}e^{-\hat{H}_m/KT}$ is the density operator of m-th molecule in the thermal equilibrium state.

This is followed by a general representation of the polarization in the quantum mechanical paradigm. It is assumed that the medium studied is small enough that the spatial variation of the optical electric field $\boldsymbol{E}(t)$ within the volume V can be ignored. In addition, the effects caused by the magnetic field associated with the optical electric field are not considered. Suppose that V contains N charged particles (electrons and ions), and q_l and r_l respectively represent the charge of the i-th particle and its position vector, then the dipole moment of the charged particle system is:

$$R = \sum_l q_l r_l \tag{2.2.11}$$

Let the macroscopic polarization of the medium be P(t). By definition, P(t) is the expected value of the dipole moment operator per unit volume.

$$P(t) = \frac{1}{V} < \hat{R} >= \frac{1}{V} \operatorname{tr}\{\hat{\rho}\hat{R}\} \tag{2.2.12}$$

where the $\operatorname{tr}\{\hat{\rho}\hat{R}\}$ is the trace of the $\{\hat{\rho}\hat{R}\}$ matrix. Since the density matrix is perturbed, there will be a technical form like the polarization of each order. Therefore, in combination with the above formula, each order polarization can be written as follows.

$$\begin{aligned} P^0 &= V^{-1} \operatorname{tr}\left\{\hat{\rho}_0 \hat{R}\right\} \\ P^{(1)} &= V^{-1} \operatorname{tr}\left\{\hat{\rho}_1(t) \hat{R}\right\} \\ P^{(2)} &= V^{-1} \operatorname{tr}\left\{\hat{\rho}_2(t) \hat{R}\right\} \\ &\vdots \\ P^{(n)} &= V^{-1} \operatorname{tr}\left\{\hat{\rho}_n(t) \hat{R}\right\} \\ &\vdots \end{aligned} \tag{2.2.13}$$

In the (2.1.3) section, the formula (2.1.7) is the formula of the nth-order polarization in the classical physics paradigm. This formula can now be rewritten in the form of a density matrix. Combining the last row in Eq. (2.2.13) with the formula (2.1.7) gives the formula for the n-th order polarizability in the quantum mechanical paradigm:

$$\begin{aligned} \chi^{(r)}_{\mu\alpha_1\alpha_2\cdots\alpha_r}(\omega_1, \omega_2, \omega_r) = \tfrac{1}{\varepsilon_0} V^{-1} (ih)^{-r} \tfrac{\hat{S}}{r!} \int_{-\infty}^{0} dt_1 \int_{-\infty}^{t_1} dt_2 \cdots \int_{-\infty}^{t_r-1} dt_r \\ \times \operatorname{tr}\left\{\hat{\rho}_0 \left[\cdots\left[\left[\hat{R}_\mu, \hat{R}^I_{\alpha_1}(t_1)\right], R^I_{\alpha_2}(t_2)\right], \hat{R}^I_{\alpha_r}(t_1)\right]\right\} e^{-i\sum_{m=1}^{r} \omega_m t_m} \end{aligned} \tag{2.2.14}$$

If the electric dipole moment operator in the equation is expressed by the electric dipole moment operator of a single molecule, then no matter which order the polarization tensor element represents, the trace has the following form:

$$P(t) = \operatorname{tr}\left\{\hat{\rho}_0 \sum_m \hat{C}_m\right\} \tag{2.2.15}$$

where the C_m in the formula in the formula represents multiple transpositions with the mth molecular electric dipole moment operator.

$$\hat{C}_m = \left[\cdots\left[\left[\hat{R}^I_{m\mu}, \hat{R}^I_{m\alpha_1}(t_1)\right], \hat{R}^I_{m\alpha_2}(t_2)\right], \ldots, \hat{R}^I_{m\alpha_r}(t_r)\right] \tag{2.2.16}$$

For example, the representation of a single molecular trace tr in a first-order polarizability tensor involves a specific form of the density matrix. In the energy representation, the operators of all functions of the Hamiltonian $\hat{H}_0$ are diagonalized:

$$
\begin{aligned}
[\hat{\rho}_0]_{ba} = \int u^*(b,r)\hat{\rho}_0 u(a,r)d\tau = A^{1/M}e^{-E_a/KT}\delta_{ab} = \rho^0_{aa}\delta_{ab} \\
\left[\hat{U}_0(t)\right]_{ba} = \int u^*(a,r)\hat{U}_0 u(a,r)d\tau = e^{-iE_b t/h}\delta_{ab} \\
\left[\hat{U}_0(t)\right]_{ba} = e^{-iEt/h}\delta_{ab} \\
\left[\hat{R}^I_a(t)\right]_{ba} = \left[\hat{U}_0(-t)\hat{R}_a\hat{U}_0(t)\right]_{ab} = e^{i\omega_{ab}t}R^a_{ab}
\end{aligned}
\tag{2.2.17}
$$

Combined with the previous derivation, the formula for the first-order polarizability tensor element can be derived:

$$
\begin{aligned}
\chi^{(1)}_{\mu a}(\omega) = -\frac{n}{\varepsilon_0}(ih)^{-1}\int_{-\infty}^{0} dt_1 \Bigg\{ \sum_{a,b} \rho^0_{aa} R^{\mu}_{ab} R^{\alpha}_{ba} e^{-i(\omega+\omega_{ab})t_1} \\
- \sum_{a,b} \rho^0_{aa} R^{\alpha}_{ab} R^{\mu}_{ba} e^{-i(\omega-\omega_{ab})t_1} \Bigg\}
\end{aligned}
\tag{2.2.18}
$$

where $n = M/V$ is the molecular number density. By integrating the above equation, we can see that when ω is a real number, the integral does not converge. Only when the frequency ω takes values in the upper half plane of the complex frequency plane, the integral is convergent. For the integral of Eq. (2.2.18), the formula for the first-order polarizability in the quantum mechanical paradigm is obtained:

$$
\chi^{(1)}_{\mu a}(\omega) = -\frac{n}{\varepsilon_0 h}\sum_{a,b}\rho^0_{aa}\left[\frac{R^{\mu}_{ab}R^{\alpha}_{ba}}{\omega+\omega_{ab}} - \frac{R^{\alpha}_{ab}R^{\mu}_{ba}}{\omega-\omega_{ab}}\right]
\tag{2.2.19}
$$

2.2.4 The Tensor and Properties of the Polarizability of the Molecular System with Inter-molecular Interaction

Suppose we discuss two energy states a and b of a single molecule, and the corresponding two eigenstate energies are $E_a = \omega_a$ and $E_b = \omega_b$, respectively. If we consider weak interactions between molecules, we will have an indeterminate amount of single-molecule energy level, which is the same magnitude as the energy of interaction between molecules. The previous section gives the representation of the first-order polarizability tensor element of an independent molecule:

$$
\chi^{(1)}_{\mu\alpha} = \frac{n}{\varepsilon_0 h}\sum_{a,b}\rho^0_{aa}\left[\frac{R^{\mu}_{ab}R^{\alpha}_{ba}}{\omega_{ba}-\omega} + \frac{R^{\alpha}_{ab}R^{\mu}_{bc}}{\omega_{ba}-\omega}\right]
\tag{2.2.20}
$$

If we consider weak interactions between molecules, a damping term ($\pm i\Gamma_{ab}$) should be introduced in the denominator, and considering the analytical requirements of the polarizability tensor, the pole on the real axis in the above equation should be moved to The lower half of the complex frequency plane. Thus, the first-order polarizability tensor element expression considering the intermolecular weak interaction medium can be obtained as:

$$\chi_{\mu\alpha}^{(1)} = \frac{n}{\varepsilon_0 h}\sum_{a,b}\rho_{aa}^0\left[\frac{R_{ab}^{\mu}R_{ba}^{\alpha}}{\omega_{ba}-\omega-i\Gamma_{ba}}+\frac{R_{ab}^{\alpha}R_{bc}^{\mu}}{\omega_{ba}+\omega+i\Gamma_{ba}}\right] \tag{2.2.21}$$

The power relationship of a unit volume of medium absorbed by linear polarization proves that Γ_{ab} is the resonance linewidth of the transition between states a and b. Assume that the medium is subjected to an optical electric field:

$$E(t) = E(\omega)\mathrm{e}^{-i\omega t}+E^*(\omega)\mathrm{e}^{i\omega t} \tag{2.2.22}$$

When the optical frequency ω is close to the transition frequency between a pair of energy states 1 and 2, the linearly polarized energy absorbed per unit volume of the medium is:

$$\begin{aligned} W &= 2\,\mathrm{Re}\left[i\omega E\cdot\left(P^{(1)}\right)^*\right] \\ &= 2\,\mathrm{Re}\left[-i\omega E^*\cdot P^{(1)}\right] \\ &= 2\,\mathrm{Re}\left[-i\omega\varepsilon_0\chi_{\mu\alpha}^{(1)}(\omega)E_{\mu}^*E_{\alpha}\right] \end{aligned} \tag{2.2.23}$$

When the intermolecular interaction is considered in the polarizability, the damping term of iGma appears on the denominator. At this time, the time reversal symmetry of the polarizability is no longer satisfied (2.1.19). However, regardless of whether the frequency is complex or real, the first-order polarizability tensor $\chi^{(1)}(\omega)$ is a symmetric tensor:

$$\begin{aligned} \chi_{\alpha\mu}^{(1)}(\omega) &= \frac{n}{\varepsilon_0 h}\sum_{a,b}\rho_{aa}^0\left[\frac{R_{ab}^{\mu}R_{ba}^{\alpha}}{\omega_{ba}-\omega-i\Gamma_{ba}}+\frac{R_{ab}^{\alpha}R_{ba}^{\mu}}{\omega_{ba}+\omega+i\Gamma_{ba}}\right] \\ &= \frac{n}{\varepsilon_0 h}\sum_{a,b}\rho_{aa}^0\left[\frac{R_{ab}^{\mu}R_{ba}^{\alpha}}{\omega_{ba}-\omega-i\Gamma_{ba}}+\frac{R_{ab}^{\alpha}R_{ba}^{\mu}}{\omega_{ba}+\omega+i\Gamma_{ba}}\right] \\ &= \chi_{\mu\alpha}^{(1)}(\omega) \end{aligned} \tag{2.2.24}$$

The correction for the expression of the high-order polarizability tensor element is similar to the correction method for the first-order polarizability tensor. When considering weak interaction between molecules, the transition frequency ω_{ab} in the formula may be replaced by $\pm i\Gamma_{ab}$ or the like. Whether it is replaced by $\omega_{ba}+i\Gamma_{ab}$ or $\omega_{ab}-i\Gamma_{ba}$, it is determined by the causality condition that the pole of the polarizability tensor should appear in the lower half plane of the complex frequency plane. Therefore, the expressions of the second-order and third-order polarizability tensor elements are:

$$\chi^{(2)}_{\mu\alpha\beta}(\omega_1,\omega_2)=\frac{\hat{S}}{2!}\frac{n}{\varepsilon_0 h^2}\sum_{a,b,c}\rho^0_{aa}\left\{\frac{R^{\mu}_{ab}R^{\alpha}_{bc}R^{\beta}_{ca}}{(\omega_{ba}-\omega_1-\omega_2-i\Gamma_{ba})(\omega_{ca}-\omega_2-i\Gamma_{ca})}\right.$$
$$+\frac{R^{\mu}_{ab}R^{\alpha}_{bc}R^{\beta}_{ca}}{(\omega_{ba}+\omega_1+i\Gamma_{ba})(\omega_{ca}-\omega_2-i\Gamma_{ca})}$$
$$\left.+\frac{R^{\mu}_{ab}R^{\alpha}_{bc}R^{\beta}_{ca}}{(\omega_{ba}+\omega_1+i\Gamma_{ba})(\omega_{ca}+\omega_1+\omega_2+iI_{ca})}\right\} \tag{2.2.25}$$

$$\chi^{(3)}_{\mu\alpha\beta\gamma}(\omega_1,\omega_2,\omega_3)=\frac{\hat{S}}{3!}\frac{n}{\varepsilon_0\hbar^3}\sum_{a,b,c,d}\rho^0_{aa}$$
$$\times\left\{\frac{R^{\mu}_{ab}R^{\alpha}_{bc}R^{\beta}_{cd}R^{\gamma}_{da}}{(\omega_{ba}-\omega_1-\omega_2-\omega_3-i\Gamma_{ba})(\omega_{ca}-\omega_2-\omega_3-i\Gamma_{ca})(\omega_{da}-\omega_3-i\Gamma_{da})}\right.$$
$$+\frac{R^{\mu}_{ab}R^{\alpha}_{bc}R^{\beta}_{cd}R^{\gamma}_{da}}{(\omega_{ba}+\omega_1+iI_{ba})(\omega_{ca}-\omega_2-\omega_3-iI_{ca})(\omega_{da}-\omega_3-i\Gamma_{da})}$$
$$+\frac{R^{\mu}_{ab}R^{\alpha}_{bc}R^{\beta}_{cd}R^{\gamma}_{da}}{(\omega_{ba}+\omega_1+i\Gamma_{ba})(\omega_{ca}+\omega_1+\omega_2+iI_{ca})(\omega_{da}-\omega_3-i\Gamma_{da})}$$
$$\left.+\frac{R^{\mu}_{ab}R^{\alpha}_{bc}R^{\beta}_{cd}R^{\gamma}_{da}}{(\omega_{ba}+\omega_1+i\Gamma_{ba})(\omega_{ca}+\omega_1+\omega_2+i\Gamma_{ca})(\omega_{da}+\omega_1+\omega_2+\omega_3+i\Gamma_{da})}\right\} \tag{2.2.26}$$

2.2.5 *Resonance Enhanced Polarizability*

In the equation of the first-order polarizability tensor element expression (2.2.18), the summation of a, b is first considered. Assume that the medium is a two-level system, the low energy level is marked with o, the high energy level is marked with t and the intrinsic transition resonance frequency is ω_{to}. In the summation process, a, b can be o and t, respectively.

$$\chi^{(1)}_{\mu\alpha}(-\omega,\omega)=\frac{n}{\varepsilon_0 h}\left[\rho^0_{oo}\left(\frac{R^{\mu}_{ot}R^{\alpha}_{to}}{\omega_{to}-\omega-i\Gamma_{to}}+\frac{R^{\alpha}_{ot}R^{\mu}_{to}}{\omega_{to}+\omega+i\Gamma_{to}}\right)\right.$$
$$\left.+\rho^0_{tt}\left(\frac{R^{\mu}_{ot}R^{\alpha}_{ot}}{\omega_{ot}-\omega-i\Gamma_{ot}}+\frac{R^{\alpha}_{to}R^{\mu}_{ot}}{\omega_{ot}+\omega+i\Gamma_{ot}}\right)\right] \tag{2.2.27}$$

Assuming the incident light frequency $\omega \approx \omega_{to}$, the first and fourth terms in the above equation are small because of the denominator value, and the fractional values are large, while the second and third terms are large due to the denominator value and can be ignored. Thus, the resonance polarizability

$$\begin{aligned}\chi_{\mu\alpha}^{(1)}(-\omega,\omega) &= \frac{n}{\varepsilon_0 h}\left(\rho_{oo}^0 - \rho_{tt}^0\right)\frac{R_{ot}^{u} R_{to}^{\alpha}}{\omega_{to} - \omega - i\Gamma_{to}} \\ &= \frac{n}{\varepsilon_0 h}\left(n_0 - n_t\right)\frac{R_{ot}^{u} R_{to}^{\alpha}}{\omega_{to} - \omega - i\Gamma_{to}}\end{aligned} \tag{2.2.28}$$

If the $\hat{S}$ operator in (2.2.25) is expanded. When the incident light is ω_1 and ω_2, the polarizability of a new coherent light with a frequency of $\omega_1 + \omega_2$ can be expressed as:

$$\begin{aligned}\chi_{\mu\alpha\beta}^{(2)}\left(-\left(\omega_1+\omega_2\right),\omega_1,\omega_2\right) = &\frac{n}{2\varepsilon_0 h^2}\sum_{a,b,c}\rho_{aa}^0 \\ &\times\left[\frac{R_{ab}^{\mu} R_{bc}^{\alpha} R_{ca}^{\beta}}{\left(\omega_{ba}-\omega_1-\omega_2-i\Gamma_{ba}\right)\left(\omega_{ca}-\omega_2-i\Gamma_{ca}\right)}\right. \\ &+\frac{R_{ab}^{\mu} R_{bc}^{\beta} R_{ca}^{\alpha}}{\left(\omega_{ba}-\omega_2-\omega_1-i\Gamma_{ba}\right)\left(\omega_{ca}-\omega_1-i\Gamma_{ca}\right)} \\ &+\frac{R_{ab}^{\alpha} R_{bc}^{\mu} R_{ca}^{\beta}}{\left(\omega_{ba}+\omega_1+i\Gamma_{ba}\right)\left(\omega_{ca}-\omega_2-i\Gamma_{ca}\right)} \\ &+\frac{R_{ab}^{\beta} R_{bc}^{\mu} R_{ca}^{\alpha}}{\left(\omega_{ba}+\omega_2+i\Gamma_{ba}\right)\left(\omega_{ca}-\omega_1-i\Gamma_{ca}\right)} \\ &+\frac{R_{ab}^{\alpha} R_{bc}^{\beta} R_{ca}^{\mu}}{\left(\omega_{ba}+\omega_1+i\Gamma_{ba}\right)\left(\omega_{ca}+\omega_1+\omega_2+i\Gamma_{ca}\right)} \\ &\left.+\frac{R_{ab}^{\beta} R_{bc}^{\alpha} R_{ca}^{\mu}}{\left(\omega_{ba}+\omega_2+i\Gamma_{ba}\right)\left(\omega_{ca}+\omega_2+\omega_1+i\Gamma_{ca}\right)}\right]\end{aligned} \tag{2.2.29}$$

Now consider the simultaneous injection of three frequencies ω_1, ω_2 and ω_3 to produce a four-wave mixing process of the fourth frequency $(\omega_1 + \omega_2 + \omega_3)$. Let the sum of the two frequencies ω_2, ω_3 and $(\omega_2 + \omega_3)$ resonate with an intrinsic transition frequency ω_{to} of the medium, that is, the condition $\omega_{to} - (\omega_2 + \omega_3) \approx 0$ is satisfied. In the summation operation of the third-order polarizability tensor element expression (2.2.26), a and c are respectively equal to o, t, where the first and second terms have resonance enhancement contributions, and the third and fourth terms There is also a resonance enhancement contribution after performing the intrinsic transaction. If the non-resonance enhancement is ignored, the resonance enhancement third-order polarizability is:

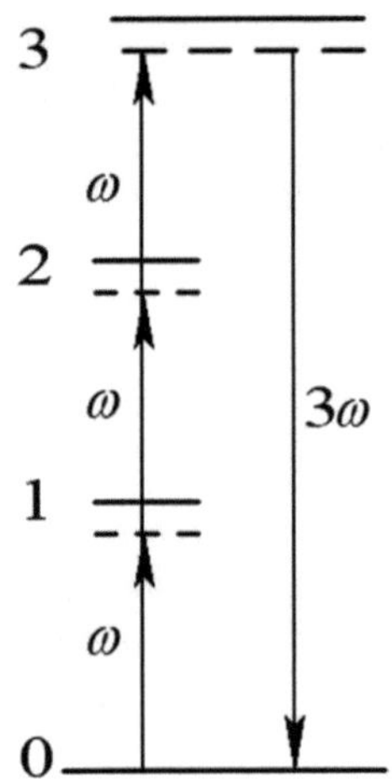

Fig. 2.1 Third harmonic generated by simple four-level system

$$\chi^{(3)}_{\mu\alpha\beta\gamma}\left(-\left(\omega_1+\omega_2+\omega_3\right),\omega_1,\omega_2,\omega_3\right)=\frac{n}{6\varepsilon_0 h^3}\frac{\rho^0_{oo}-\rho^0_{tt}}{\omega_{to}-(\omega_2+\omega_3)-i\Gamma_{to}}$$
$$\times\sum_b\left[\frac{R^{\mu}_{tb}R^{\alpha}_{bt}}{\omega_{bo}-\omega_1-\omega_2-\omega_3}+\frac{R^{\alpha}_{tb}R^{\mu}_{bt}}{\omega_{bo}+\omega_1}\right]$$
$$\times\sum_b\left[\frac{R^{\beta}_{tb}R^{\gamma}_{bo}}{\omega_{bo}-\omega_3}+\frac{R^{\gamma}_{tb}R^{\beta}_{bo}}{\omega_{bo}-\omega_2}\right] \tag{2.2.30}$$

Figure 2.1 shows the four-wave mixing process of a simple four-level system. When the intermediate level is exactly equal to the initial state and the final state, the process becomes a resonance process. The dotted level in the figure represents the corrected energy level after considering the intermolecular interaction.

Now consider ω_1, ω_2, and ω_3 to produce a four-wave mixing process of $(\omega_1-\omega_2+\omega_3)$, where the difference between the two incident light frequencies exactly coincides with an intrinsic transition frequency of the medium, such as $\omega_{to}\approx\omega_2-\omega_3$. At this point, starting from Eq. (2.2.26), the third-order polarizability tensor element describing the process can be derived as:

$$\chi^{(3)}_{\mu\alpha\beta\gamma}\left(-\left(\omega_1-\omega_2+\omega_3\right),\omega_1-\omega_2,\omega_3\right)=\frac{n}{3!\varepsilon_0 h^3}\frac{\rho^0_{oo}-\rho^0_{tt}}{\left[\omega_{to}-(\omega_2-\omega_3)+i\Gamma_{to}\right]}$$
$$\times\sum_b\left[\frac{R^{\mu}_{ob}R^{\alpha}_{bt}}{\omega_{bo}+\omega_1-\omega_2+\omega_3}+\frac{R^{\alpha}_{ob}R^{\mu}_{bt}}{\omega_{bo}-\omega_1}\right]\sum_b\left[\frac{R^{\beta}_{ob}R^{\gamma}_{bt}}{\omega_{bo}-\omega_2}+\frac{R^{\gamma}_{ob}R^{\beta}_{bt}}{\omega_{bo}-\omega_3}\right] \tag{2.2.31}$$

In actual work, ω_{to} is often chosen as the Raman transition frequency of the medium. At this time, a so-called Raman resonance-enhanced four-wave mixing process will occur, and various Raman resonance four-wave mixing can be described by using Eq. (2.2.31) process. Normally, two different sources of monochromatic light are incident, one of which is called pump light (ω_p) and the other weaker is called signal light (ω_s). The following four effects:

Raman Gain Effect

This effect is described by $\chi^{(3)}\left(-\omega_{\text{s}}, \omega_{\text{p}}, -\omega_{\text{p}}, \omega_{\text{s}}\right)$, where $\omega_{\text{p}} > \omega_{\text{s}}$, the incident signal gain is obtained at the frequency of $\omega = \omega_{\text{p}} - \left(\omega_{\text{p}} - \omega_{\text{s}}\right) = \omega_{\text{p}} - \omega_{t\varrho} = \omega_{\text{s}}$, which constitutes The basis of Raman gain spectroscopy.

Anti-Stokes Raman Attenuation Effect

This effect is described by

$$\chi^{(3)}\left(-\omega_{\text{s}}, \omega_{\text{p}}, \omega_{\text{s}}, -\omega_{\text{p}}\right) = [\chi^{(3)}\left(\omega_{\text{s}}, -\omega_{\text{p}}, -\omega_{\text{s}}, \omega_{\text{p}}\right)]^{*},$$

where $\omega_{\text{s}} > \omega_{\text{p}}$, at $\omega = \omega_{\text{p}} - \left(\omega_{\text{p}} - \omega_{\text{s}}\right) = \omega_{\text{p}} + \omega_{to} = \omega_{\text{s}}$ The attenuation of the incident signal occurs at the frequency and becomes the basis of Anti-Stokes spectroscopy.

Coherent Stokes Light Generation

This effect is described by

$$\chi^{(3)}\left(-\left(2\omega_{\text{s}} - \omega_{\text{p}}\right),\quad \omega_{\text{s}}, -\omega_{\text{p}}, \omega_{\text{s}}\right)$$

where $\omega_{\text{p}} > \omega_{\text{s}}$, which produces a spatial orientation at the frequency $\omega = \omega_{\text{s}} - \left(\omega_{\text{p}} - \omega_{\text{s}}\right) = \omega_{\text{s}} - \omega_{\text{to}}$ The new Stokes frequency shifts the beam.

Coherent Anti-Stokes Light Generation

This effect is described by

$$\chi^{(3)}\left(-\left(2\omega_{\text{p}} - \omega_{\text{s}}\right), \omega_{\text{p}}, \omega_{\text{p}}, -\omega_{\text{s}}\right) = \left[\chi^{(3)}\left(\left(2\omega_{\text{p}} - \omega_{\text{s}}\right), -\omega_{\text{p}}, -\omega_{\text{p}}, \omega_{\text{s}}\right)\right]^{*}$$

where $\omega_{\text{p}} > \omega_{\text{s}}$, a new anti-Stokes frequency-shifted beam that produces a spatial orientation at the frequency of $\omega = \omega_{\text{p}} + \left(\omega_{\text{p}} - \omega_{\text{s}}\right) = \omega_{\text{p}} + \omega_{to}$, which is the basis of coherent anti-Stokes Raman spectroscopy.

2.2.6 Calculation Method of Nonlinear Polarizability by Higher-Order Derivative

There are many calculation methods for polarizability in modern quantum chemistry. There are also some commercial software that already have mature algorithms for calculating the polarizability. Gaussian [59] is one of the most popular quantum chemistry software. The polar keyword in Gaussian is specifically used to calculate the polarizability α and the first hyperpolarizability β (G09 also supports polar = gamma to calculate the second hyperpolarizability γ from the later stage). The algorithm is based on the energy expansion method for the field of the field to calculate the polarizability:

$$E(\mathbf{F}) = E(\mathbf{0}) + \left.\frac{\partial E}{\partial \mathbf{F}}\right|_{\mathbf{F}=0} \mathbf{F} + \frac{1}{2}\left.\frac{\partial^2 E}{\partial \mathbf{F}^2}\right|_{\mathbf{F}=0} \mathbf{F}^2 + \frac{1}{6}\left.\frac{\partial^3 E}{\partial \mathbf{F}^3}\right|_{\mathbf{F}=0} \mathbf{F}^3 + \frac{1}{24}\left.\frac{\partial^4 E}{\partial \mathbf{F}^4}\right|_{\mathbf{F}=0} \mathbf{F}^4 + \ldots$$
$$= E(\mathbf{0}) - \mu_0 \mathbf{F} - \frac{1}{2}a\mathbf{F}^2 - \frac{1}{6}\beta\mathbf{F}^3 - \frac{1}{24}\gamma\mathbf{F}^4 - \frac{1}{120}\mathbf{F}^5 - \frac{1}{720}\varepsilon\mathbf{F}^6 \ldots \quad (2.2.32)$$

According to this series expansion, the polarization of each order can be obtained:

$$\mu_0 = -\left.\frac{\partial E}{\partial \mathbf{F}}\right|_{\mathbf{r}=0} \quad \alpha = -\left.\frac{\partial^2 E}{\partial \mathbf{F}^2}\right|_{\mathbf{r}=0} \quad \beta = -\left.\frac{\partial^3 E}{\partial \mathbf{F}^3}\right|_{\mathbf{r}=0} \quad \gamma = -\left.\frac{\partial^4 E}{\partial \mathbf{F}^4}\right|_{\mathbf{r}=0} \quad (2.2.33)$$

μ_0 is the permanent dipole moment vector of the numerator (no dipole moment in the external field). α is the polarizability of the molecule, also called the linear optical coefficient, which is a 3*3 s-order tensor. β is the first hyperpolarizability, also known as the second-order nonlinear optical (NLO) coefficient of the molecule, which is a 3*3*3 third-order tensor. γ is the second hyperpolarizability, also known as the third-order NLO coefficient of the molecule, which is relatively less studied. The higher σ is rarely discussed. The (super) polarizability value is related to the electric field frequency. The case where the external field frequency is 0 is called the static (super) polarizability; if the external field frequency is not 0, such as light of a certain frequency, the corresponding is dynamic (super Polarizability, or frequency (super) polarizability.

Gaussian's polar keyword seeks these quantities through the derivative method, and the principle is clear. As can be seen from the above formula, the polarization rate should be such that the energy makes two derivatives to the external field, and the first hyperpolarization rate is required to be three derivatives.

The derivative of energy is divided into three cases:

(1) Analyze the derivative. This method is fast and accurate, but programming is difficult, especially for advanced post-HF. In this way, the derivative of the molecular orbital coefficient to the external field is required to obtain the energy derivative, which is achieved by solving the CPHF equation. And the dynamic (super)

polarizability can be obtained by using the frequency-containing CPHF equation. This way of finding (super) polarizability is also known as the CPHF method.

(2) Numerical derivatives. This way the derivative of energy is obtained by finite difference method, also known as the finite field (FF) method. Finite difference is described in all numerical algorithms. For example, the derivative of a function f at 0 can be expressed as $f(0) = [f(0.001) - f(-0.001)/(2 * 0.001)$. Finding (super) polarizability by this method is very time consuming. It is necessary to calculate many single points in different fields and different directions, and the accuracy of high-order derivatives is very poor, because the numerical derivative will be numerically every time. The cumulative error of noise increases rapidly with the increase of the order of the derivative, and the accuracy is very bad when the second derivative is used (unless it is processed by special methods such as additional correction and extrapolation). In addition, the dynamic (super) polarizability cannot be obtained by numerical derivatives.

(3) Analyze + value mixed derivative. This is based on the low-order analytical derivative, and then the high-order derivative is obtained by one or more finite difference to obtain the (super) polarizability. Accuracy and speed are between the full analytical derivative and the full numerical derivative.

G09 is already very good at supporting analytical derivatives. For HF, DFT, and semi-empirical methods, it can support third-order analytical derivatives, so it can fully and statically obtain static and frequency-containing α and β (while other quantization programs are very It is difficult to support such high-order derivatives for HF/DFT/semi-experience, and usually support second order). For such methods that support third-order analytical derivatives, the use of the polar keyword directly gives the results of α and β. If CPHF = RdFreq is written at the same time, and the external field frequency ω is written in a blank line at the end of the input file, such as 0.05 0.12 0.3 (the static situation will always be calculated, so it is not necessary to write 0.0), the frequency polarization ratio will be calculated. $\alpha(-\omega; \omega)$ and the frequency first polarization rate $\beta(-\omega; \omega, 0)$. If the frequency-bearing first polarizability to be calculated is the SHG form, that is, $\beta(-\omega; \omega, 0)$, write polar = DCSHG, and CPHF = RdFreq can be ignored or not, and $\beta(-\omega; \omega, 0)$ will also be calculated.

For G9 such as MP2, only the method of second-order analytical derivative can be supported, and α can be calculated fully analytically, but if you want to calculate β, you have to do a finite difference again. In G09, directly writing polar will only calculate α. If you want to get β, you have to write polar = Cubic. The third derivative is automatically obtained based on the analytical second derivative.

For CISD, QCISD, CCSD, MP3, MP4 (SDQ) and other G09 only supports the first-order analytical derivative method. Directly writing the polar keyword will make a second derivative based on the analytical first-order derivative to obtain the second derivative, ie α. If you want to get β, you have to write polar = DoubleNumer (equivalent to polar = enonly), and you will get the third derivative by making two finite differences based on the first derivative.

For CCSD(T), QCISD (T), MP4 (SDTQ), MP5, etc., only the method of calculating the single point energy is supported. Writing the polar keyword will make two finite differences in energy and thus obtain α. There is no way for them to give β directly in G09.

Note that for methods other than HF, DFT, and semi-empirical, G09 cannot give a frequency (super) polarizability.

If you want to obtain a higher order hyperpolarizability for higher order derivatives, such as calculating the second hyperpolarizability γ under DFT, you need to make a finite difference based on its analytical third derivative. Starting from G09 D.01, for the method supporting the third-order analytical derivative, static = gamma can be used to calculate static γ and dynamic $\gamma(-2\omega; \omega, \omega, 0)$, $\gamma(-\omega; \omega, 0, 0)$, the finite difference is automatically made based on the analytical third derivative to get the required fourth derivative. At the end, the external field frequency should be written in a blank line.

The following is an example of an input file for calculating the polarizability in Gaussian. The line starting with "#p" is the line defining the keyword in Gaussian. PBE1PBE is a functional of density functional theory. "PBE1PBE" is a special notation in Gaussian, which represents the hybrid functional PBE0. "aug-cc-pVTZ" is an abbreviation for the basis function group. "aug" is the diffusion function identifier. A diffusion function must be included in the task of calculating the polarizability. The problem of the dispersion function will be explained in detail later. "polar" is a keyword for calculating the polarizability. This keyword can be followed by different options, such as α, β and γ, etc. to calculate different polarizabilities. "CPHF = RdFreq" is a keyword that controls the incident wavelength at the end of the read file. This keyword can be written without writing to calculate the static polarizability. If the wavelength of the incident light (the energy of the atomic unit) is set at the end of the file, the dynamic polarizability at this energy can be calculated.

```
#p PBE1PBE/aug-cc-pVTZ polar CPHF=RdFreq

polarizability calculated method

0 1                          // Charge and Multiplicity
{molecular structure}        // XYZ

0.07 0.1                     // incident energy
```

The importance of the diffusion function is then introduced. The dipole moment, the polarizability, and the first hyperpolarizability are the first, second, and third derivatives of the energy external electric field, respectively. As the derivative order increases, the requirements for the dispersion function become higher and higher. The most cost-effective base groups for this problem are from small to large, ZPOL, jul-cc-pVDZ, aug-cc-pVDZ, POL, aug-cc-pVTZ (-f, -d), LPol-ds. Among them, aug-cc-pVTZ (-f, -d) is obtained by cutting off the f-polarization of the heavy atom of aug-cc-pVTZ and the d-polarization of the light atom. In addition, although the base group of the def2 series base group with the D suffix such as def2-SVPD is also optimized for the polarization calculation, it is not as good as the above base group in my test, but it can be used if desired. If the potential is used, it is recommended to use the LFK potential base group, which is modified on the basis of the SBKJC potential base group, so that the calculation accuracy of the polarizability is similar to that of the all-electronic POL base group (http://sobereva.com/336).

Adding a dispersion function will increase the completeness of the basis function. In principle, it seems that it should always be beneficial to the result, but in the actual calculation, it will lead to the following problems:

1. The amount of calculation has soared. This is well known, especially the cc-pVnZ series, which often does not move after adding aug-, and the fourth section will discuss how to solve this difficulty.
2. SCF is often more difficult to converge than when it is not added after adding the dispersion function.
3. The chemical significance of the diffusion function is very poor, and the lack of correspondence with the atomic orbital will have a serious adverse effect on the wave function analysis method under the Hilbert space. For example, the Mulliken charge will become extremely bad, and the Mayer key level will be quite unreliable. The reason is not difficult to understand. For example, a large number of diffusion functions of A atoms extend into the space of B atoms, so some of the electron distribution near B will be described by these dispersion functions, then Mulliken population analysis will put a lot of B should belong to B. The electron is assigned to the A atom, causing the charge of A to be too negative and B to be too positive.
4. The chemical significance of the virtual orbit (ie, the non-occupied orbit) becomes more ambiguous. Especially in the Hartree-Fock calculation under the dispersion function, the spatial distribution of the virtual orbit is often very wide, which makes the theoretical analysis of the frontier track completely unsuitable.
5. The numerical problem caused by the linear dependence of the basis function. However, in the general quantization procedure, the eigenvalues of the basis function overlap matrix are automatically tested to properly cut off some basis functions to solve this problem.
6. The symmetry of the architecture and wave function may be degraded due to numerical accuracy problems. For example, if you optimize a system with symmetry, the initial structure program can judge the actual point group, but after optimization, the program can only judge the lower order group. If you currently use the dispersion function, then 80% is caused by the dispersion function, and the dispersion function can often maintain the symmetry.
7. If the original basis set is not complete, but the excessive dispersion function is added, the effect that should be represented by the valence layer basis function will instead be represented by the dispersion function, which may cause unreasonable research in some problems. As a result, this actually belongs to the intramolecular BSSE category. An example is JACS, 128, 9342 found that the stable structure of

benzene calculated by HF combined with some diffuse versions of the Pople basis set (such as 6–31++G**) turned out to be curved, or a stable plane. The structure has a virtual frequency. This is because there are no higher angular momentum basis functions (especially f) that are more dependent on the post-HF calculation in these Pople basis sets, and the more diffuse s and p basis functions that extend over the past provide a higher angular momentum basis for carbon. The effect of the function causes the bending of the structure.

Let me talk about the general characteristics of the dispersion function. The index of the dispersion function for each angular momentum is less than a multiple of the minimum exponent of the other equivalent angular momentum functions in the base set. The number of dispersion functions in the various basis sets and the angular momentum involved are different. The dispersion function is generally non-shrinking.

Since the dispersion function is important in many situations, most of the mainstream base groups have versions with dispersion functions. Some versions with dispersion functions were created by the authors of the original base group, and others were proposed by other researchers. Here are some common ones:

1. Pople series base group: only add a layer of sp (that is, a layer of the same index and a layer of p) to the heavy atom. The dispersion function adds a plus sign to the front of the G, such as 6–31+G*; Adding a s dispersion function to the hydrogen and helium atoms adds a plus sign to the front of the G, such as 6–311++G (2df, 2p). There are many basic groups in the Pople series, but the indices of the diffusion function are shared and not optimized separately. (It is worth mentioning that the time consumption of 6–311G* is lower than 6–31+G*. When the dispersion function does not play a key role, the former result is better than the latter.
2. Dunning-related consistency basis set (cc-pVnZ series): plus the version of the dispersion function is aug-cc-pVnZ series (aug = augmented), which is given to each angular momentum function of the corresponding cc-pVnZ basis set. A layer of diffusion function with equal angular momentum is added. For example, cc-pVTZ is 4 s, 3p, 2d, 1f for C, so the aug-version will add a layer s, a layer p, a layer d and a layer f dispersion function. While cc-pVDZ is 2 s, 1p for hydrogen, the aug-version adds a layer of s and a layer of p dispersion. The same dispersion function as aug-cc-pVnZ is added to the cc-pCVnZ and cc-pwCVnZ basis groups which describe the kernel correlation to become aug-cc-pCVnZ and aug-cc-pwCVnZ; added to cc-pVnZ suitable for DKH calculation -DK becomes aug-cc-pVnZ-DK; added to cc-pV(n+d)Z series (this set of bases is based on cc-pVnZ to add a tight d function to improve extrapolation Convergence) becomes aug-cc-pV(n+d)Z, where only the index of the d-dispersion function is re-optimized. The relevant consistency base group cc-pVnZ-PP also has a version of the dispersion function aug-cc-pVnZ-PP. The type and quantity of the diffusion function are the same as the aug-cc-pVnZ series, but the indices are re-optimized. In addition, it is also possible to add a multi-layer dispersion function to each angular momentum of the cc-pVnZ series. d-aug-cc-pVnZ and t-aug-cc-pVnZ are two and three layers of

diffusion for each angular momentum. The function is extremely expensive and is generally used for accurately calculating the excited state and hyperpolarizability of Rydberg.

3. Ahlrichs' def2-series base set: There is currently no official version of the diffusion function with this type of base set.
4. Jensen's Polarization Consistency Base Group pc-n: In JCP, 117, 9234, Jensen proposed a method to increase the dispersion function for his pc-n series basis set. The diffusion function can be added to the multi-high angle momentum as needed select. The results show that the addition of s and p dispersion can greatly improve the accuracy of the electron affinity of DFT calculation, and further improve the response property calculation results also need to add a higher angular momentum dispersion function.
5. Lanl's base group: In JPCA, 105, 8111, the author added a layer of d-polarization and a layer of p-dispersion function to the main family Lanl2DZ, named LANL2DZdp, in calculating the electron affinity, vibration frequency, and bond. The long side is much better than the Lanl2DZ. LANL2TZ+ and LANL08+ add a layer of dispersion to the LANL2TZ and LANL08 for the first-period transition metal, respectively. This is due to the fact that the transition metal of the first period filled with the d-shell is sometimes easily polarized.

Except that the index of the dispersion function of the base group such as LANL2TZ+ and LANL08+ is derived by the even-tempered method, most of the indices of the diffusion function mentioned above are derived from different ways to minimize the energy of the anion calculation. The shrinkage coefficient and index of the group remain unchanged, and the dispersion function is simply added to the original basis set). However, the calculation of a good energy dispersion function basis set is used to calculate other properties, especially the (super) polarization rate that strongly depends on the dispersion function is not necessarily good, or the cost performance is not high. In order to make the response properties such as (super) polarizability have satisfactory calculation results at lower computational cost, the diffusion function and even the entire base group are directly derived from the calculation of the optimized response properties. A few examples:

Sadlej POL: Also known as the Sadlej pVTZ base group, it has been proposed since 1988 that the parameters are obtained by optimizing the calculation of the polarizability. The size is close to cc-pVTZ, and the accuracy of polarization is close to the much more expensive aug-cc-pVTZ.

Sadlej ZPOL: In 2004, it was proposed. Simplify the POL basis set. It is suitable for calculating the dipole moment and the polarizability of a large system, and it is similar to the 6–311+G* time-consuming phase, but the accuracy of the polarizability is better than that of 6–311++G (2df, 2p).

Sadlej LPol-ds: presented in 2009. The LPol series base component is LPol-ds/dl/fs/fl, which increases in turn. LPol-ds is the smallest of them, but it is much larger than POL. The accuracy of the first hyperpolarizability is very good. It is similar to d-aug-cc-pVTZ, and the time-consuming is the lowest of the same grade. Only C, H, O, N, F are defined.

def2-SVPD, TZVPD, TZVPPD, QZVPD, QZVPPD basis set: proposed in JCP, 133, 134105, is to add the dispersion function to the def2-series SVP, TZVP, TZVPP, QZVP, QZVPP basis sets, respectively. The index of the dispersion function is obtained by optimizing the HF polarizability of the atom.

2.2.7 *Nonlinear Polarizability by Sum-Over-States (SOS) Method*

Complete State Summation (SOS) is a common method for calculating polarizability and hyperpolarizability. Multiwfn supports calculation of static/frequency-containing polarizability α, first hyperpolarizability β, second hyperpolarizability γ, and even third hyperpolarizability σ by SOS. Multiwfn is a multi-function wave function analysis software developed by Dr. Tian Lu [60]. This software has a complete function of calculating high-order polarizability using the SOS method [http://sobereva.com/232]. The energy and dipole moments of each state used in the calculation, as well as the transition dipole moment between the states, can be generated by Multiwfn based on Gaussian or ORCA CIS, TDHF, TDDFT calculation results. Although Gaussian can directly use the TD (SOS) keyword to do SOS calculation based on the results of TDHF and TDDFT, it can only give the polarization rate, which is obviously too limited, and does not support SOS calculation for the commonly used CIS. The specific principles of SOS were proposed in 1993 [61].

The formula for calculating the polarizability and the first hyperpolarizability is as follows:

$$\alpha_{AB}(-\omega;\omega)=\sum_{i\neq 0}\left[\frac{\mu_{0i}^{A}\mu_{i0}^{B}}{\Delta_i-\omega}+\frac{\mu_{0i}^{B}\mu_{i0}^{A}}{\Delta_i+\omega}\right]=\hat{P}[A(-\omega),B(\omega)]\sum_{i\neq 0}\frac{\mu_{0i}^{A}\mu_{i0}^{B}}{\Delta_i-\omega}$$

$$\beta_{ABC}(-\omega_\sigma;\omega_1,\omega_2)=\hat{P}\left[A(-\omega_\sigma),B(\omega_1),C(\omega_2)\right]\sum_{i\neq 0}\sum_{j\neq 0}\frac{\mu_{0i}^{A}\overline{\mu_{ij}^{B}}\mu_{j0}^{C}}{(\Delta_i-\omega_\sigma)(\Delta_j-\omega_2)} \tag{2.2.34}$$

where $\mu_{ij}^{A}=\langle i\left|\hat{\mu}^{d}\right|j\rangle$ $\overline{\mu_{jj}^{A}}=\mu_{jj}^{A}-\mu_{00}^{A}\delta_{ij}$ $\omega_\sigma=\sum_i\omega$. A, B, C... such labels are used to indicate the direction X, Y, Z. ω is the external field energy, and when it is 0, it corresponds to the static (super) polarizability. Addition is the summation of all excited states, and Δ represents the excitation energy of the excited state relative to the ground state. $\hat{P}$ indicates that the items in the parentheses are subject to various possible permutations. For example, for β, $\hat{P}$ has three items in square brackets, so there are $3!=6$ sorts, and the results of these six cases are added. μ_{ij}^{A} represents the component in the A direction of the transition dipole moment of the two states i and j. When $i=j$ corresponds to the dipole moment of the i-th state, μ_{00} is the dipole moment of the ground state. δ_{ij} is the kronecker symbol, 1 for $i=j$, otherwise 0.

Similarly, the formula for the second and third hyperpolarizabilities is:

$$\gamma_{ABCD}(-\omega_\sigma;\omega_1,\omega_2,\omega_3)=\hat{P}\left[A(-\omega_\sigma),B(\omega_1),C(\omega_2),D(\omega_3)\right]\left(\gamma^{1}-\gamma^{\mathrm{II}}\right)$$

$$\gamma^{1}=\sum_{i\neq 0}\sum_{j\neq 0}\sum_{k\neq 0}\frac{\mu_{0i}^{A}\overline{\mu_{ij}^{B}\mu_{jk}^{C}}\mu_{k0}^{D}}{(\Delta_i-\omega_\sigma)\left(\Delta_j-\omega_2-\omega_3\right)(\Delta_k-\omega_3)}$$

$$\gamma^{\mathrm{II}}=\sum_{i\neq 0}\sum_{j\neq 0}\frac{\mu_{0i}^{A}\mu_{0i}^{B}\mu_{0j}^{C}\mu_{j0}^{D}}{(\Delta_i-\omega_\sigma)(\Delta_i-\omega_1)\left(\Delta_j-\omega_3\right)}$$

(2.2.35)

$$\delta_{ABCDE}(-\omega_\sigma;\omega_1,\omega_2,\omega_3,\omega_4)=\hat{P}\left[A(-\omega_\sigma),B(\omega_1),C(\omega_2),D(\omega_3),E(\omega_4)\right]\left(\delta^{\mathrm{I}}-\delta^{\mathrm{II}}-\delta^{\mathrm{III}}\right)$$

$$\delta^{\mathrm{I}}=\sum_{\substack{l,j,k,\\(\neq 0)}}\frac{\mu_{0i}^{A}\overline{\mu_{ij}^{B}\mu_{jk}^{C}\mu_{kl}^{D}}\mu_{l0}^{E}}{(\Delta_i-\omega_\sigma)\left(\Delta_j-\omega_\sigma+\omega_1\right)(\Delta_k-\omega_3-\omega_4)(\Delta_l-\omega_4)}$$

$$\delta^{\mathrm{II}}=(1/2)\sum_{\substack{i,j,k\\(\neq 0)}}\frac{\mu_{0i}^{A}\mu_{i0}^{B}\mu_{0j}^{C}\overline{\mu_{jk}^{D}}\mu_{k0}^{E}}{\left(\Delta_j+\omega_2\right)(\Delta_k-\omega_4)}\left(\frac{1}{\Delta_i-\omega_\sigma}+\frac{1}{\Delta_i-\omega_1}\right)\times\left(\frac{1}{\Delta_j-\omega_3-\omega_4}+\frac{1}{\Delta_k+\omega_2+\omega_3}\right)$$

$$\delta^{\mathrm{III}}=(1/2)\sum_{\substack{i,j,k\\(\neq 0)}}\frac{\mu_{0i}^{A}\mu_{i0}^{B}\mu_{0j}^{C}\mu_{jk}^{D}\mu_{k0}^{E}}{(\Delta_i-\omega_\sigma)(\Delta_i-\omega_1)}\left[\frac{1}{\left(\Delta_j-\omega_3-\omega_4\right)(\Delta_k-\omega_4)}+\frac{1}{\left(\Delta_j+\omega_2\right)(\Delta_k+\omega_2+\omega_3)}\right]$$

(2.2.36)

It can be seen that it is not difficult to do SOS calculations. The formulas are all ready-made, as long as the excitation energy, the dipole moment of each excited state, and the transition dipole moment between the excited states are provided. These quantities can be produced by electronic excitation methods such as ZINDO, CIS, TDHF, TDDFT. CIS(D) is also possible. For SOS, the transition dipole moment is still CIS, but the excitation energy is corrected by second-order perturbation to better consider the electron correlation effect.

The entire SOS calculation process is divided into two parts: (1) electronic excitation calculation (2) cyclic accumulation according to SOS formula. For the part (2), the α and β calculations themselves take almost no time, and γ takes a little time. For σ, when considering a large number of states, it can be seen from the above equation that quadruple cycle accumulation of the excited state is required, and $3^5 = 243$ components are included (although some components are the same to avoid double counting), so σ is calculated. Still quite time consuming. For the general α, β and γ we are interested in, the entire computational cost of SOS is mainly based on (1), especially for large systems and high quality base groups. Note that (2) part of the time consumption itself and the number of basis functions are not directly related, only depends on the number of states considered, but the time consumption of (1) is directly related to the number of basis functions.

ZINDO is a semi-empirical method, which is very fast, and the SOS/ZINDO combination is very cheap and is often used to calculate the NLO properties of organic large conjugate systems. The more precise CIS/TDHF/TDDFT in principle is obviously much more time consuming. In principle, SOS should sum all states. Although it is not necessary to consider all states in actual research, it is generally necessary to calculate 40–120 states in the electronic excitation process, which is much higher than the state required to study the electronic excitation problem. The more states solved, the more time-consuming CIS/TDHF/TDDFT is. Originally, the electronic excitation calculation of this kind of ab initio calculation method is difficult to use in a large system. In addition, in order to do SOS, it is necessary to calculate so many states, and it is more accurate to calculate β, especially γ, and it needs a large diffusion function. The scale of the basic group, SOS combined with CIS/TDHF/TDDFT can be very limited. Compared to SOS, if all derivatives can be calculated analytically, it is better to use the derivative method to calculate (super) polarizability. However, supporting high-order analytical derivatives is difficult from a programming perspective. Gaussian's method of achieving third-order analytical derivatives (corresponding to β) is only HF, DFT, and semi-empirical. They do not support fourth-order analytical derivatives to produce γ. Although a static γ can be obtained by making a finite difference based on the third-order analytical derivative, it is impossible to obtain a frequency-containing γ, which necessitates the use of SOS. In addition, as long as the information required for SOS calculation is available, each time the calculation of SOS is fast, it is simply a simple cycle and addition, subtraction, multiplication, and division. Therefore, it is convenient to study the changes of α, β, and γ with frequency. A distinct advantage of SOS.

The user can write the information required for the SOS calculation generated by various quantization programs into a text file in a format similar to the following, and then read and calculate the text by the polarizability and hyperpolarizability when Multiwfn is started. The meaning of the file is easy to understand and easy to write. The following is an example of an input file that uses the Multiwfn software in conjunction with the SOS method to calculate the polarization of each order.

```
2 // Excited state
1 1.1 // The first excited state,
      //its sequence number and excitation energy (eV)
2 3.2
0 0 0.845 0.2 0.4        // 0 represents the ground state.
0 1 0.231 0.3 0.7        // The transition dipole moment from the
                         // ground state to the first excited state
0 2 0.112 0.564 0.21
1 1 0.021 0.465 0.0      // Dipole moment of the first excited state
1 2 0.001 0.3 0.11       // Transition dipole moment of the
                         // 1st to 2nd excited states
2 2 0.432 0.14 0.42      // Dipole moment of the second excited state
```

The use of Multiwfn in combination with the SOS method to calculate the polarization of each order requires a transition dipole moment from the ground state to each excited state. Therefore, Gaussian or other quantum chemistry software is needed to calculate the excited state information, such as ORCA, PSI4, etc. are supported by Multiwfn. The following is an example of Gaussian's calculation of excited state information:

```
#P ZINDO(nstates=150)/gen IOp(9/40=5)

polarizability calculated method by SOS

0 1
{Molecular structure}
```

The "ZINDO" is a semi-empirical method for quickly calculating excited state information, also known as INDO/s. Of course, TDDFT, TDHF, CIS and EOM-CCSD can also be used here. However, the calculation of the polarizability using the SOS method requires a large number of excited states, and the calculation of the post-HF method is too large, so it is rarely used in the calculation of the polarizability. The "IOp(9/40 = 5)" is the key to control the accuracy of the output electronic excitation configuration system in Gaussian, which means that all electronic excitation configuration coefficients greater than 0.00001 are output. This keyword is different in different software. In the ORCA software, this keyword becomes:

```
%tddft
    tda false
    tprint n
end
```

This means that all electronic excitation configuration coefficients greater than n are output. Note that the tddft method in ORCA uses the TDA method [62] by default. The "tda false" field is used to turn off TDA and use TDDFT. The reason why TDA cannot be used here is that the electronic excitation configuration coefficient of TDA is the value after squared, so the excitation and de-excitation cannot be effectively analyzed.

When Multiwfn software is combined with the SOS method to calculate the polarizability, the calculation of the polarization of each order can be performed at different frequencies. For example, the first hyperpolarizability $\beta(-(\omega_1 + \omega_2); \omega_1, \omega_2)$ can be calculated as the sum frequency polarization of 800 nm by inputting 0.0569, 0.0569 (atomic unit, Hartree). That is to say, as long as the molecular structure is known, the corresponding polarizability can be calculated according to the frequency of different nonlinear optical processes. This is very important and can even be used to calculate the strength of nonlinear Raman processes such as SRS, CARS.

2.3 Common Nonlinear Optical Processes

There are many nonlinear optical processes, such as Second-harmonic generation (SHG), Third-harmonic generation (THG), High-harmonic generation (HHG), Sum-frequency generation (SFG), and Difference-frequency generation (DFG). Optical parametric amplification (OPA), Optical parametric oscillation (OPO), Optical parametric generation (OPG), Optical rectification (OR), generation of quasi-static electric fields, optical Kerr effect, Cross-phase modulation (XPM), Four-wave Mixing (FWM) and Self-focusing and so on. However, there are not many phenomena that can be applied in analytical chemistry, biomedical and materials science detection and characterization. Moreover, there are fewer techniques that can be used in the field of imaging. This book is intended to introduce the principles and techniques of several types of nonlinear optical signal imaging. Other effects can be found in [63] and [https://en.wikipedia.org/wiki/Nonlinear_optics]. Therefore, in this section, only the nonlinear processes used in this book are introduced, and the basic form of the polarizability is given. This basic form can be theoretically studied according to the calculation method introduced above.

2.3.1 Second-Harmonic Generation (SHG)

Second harmonic generation (SHG, also known as frequency doubling) is a nonlinear optical process in which two photons of the same frequency interact with a nonlinear material, are "combined" and produce twice as many initial photons. The new photon of energy (equivalently), twice the frequency and half the wavelength). This is a special case of sum and frequency generation. The second-order nonlinear sensitivity of the medium characterizes its tendency to cause SHG. As with other even order nonlinear optical phenomena, second harmonic generation is not allowed in media with inversion symmetry [63]. In some cases, almost 100% of the light energy can be converted to the second harmonic frequency. These conditions typically involve a strong pulsed laser beam that passes through a large crystal and is carefully aligned to obtain a phase match. In other cases, such as a second harmonic imaging microscope, only a small fraction of the light energy is converted to a second harmonic - but this light can still be detected by means of an optical filter.

Schematic diagram of SHG in Fig. 2.2. Therefore, its polarizability can be expressed as:

$$\chi^{(2)} = \beta(-2\omega_1; \omega_1, \omega_1) \tag{2.3.1}$$

Since the medium with reverse symmetry is prohibited from producing second harmonic light, the surface and interface are interesting topics for studying SHG. In fact, second harmonic generation and sum frequency generation are distinguished from a large number of signals, implicitly marking them as surface specific techniques. In 1982, TF Heinz and YR Shen demonstrated for the first time that SHG

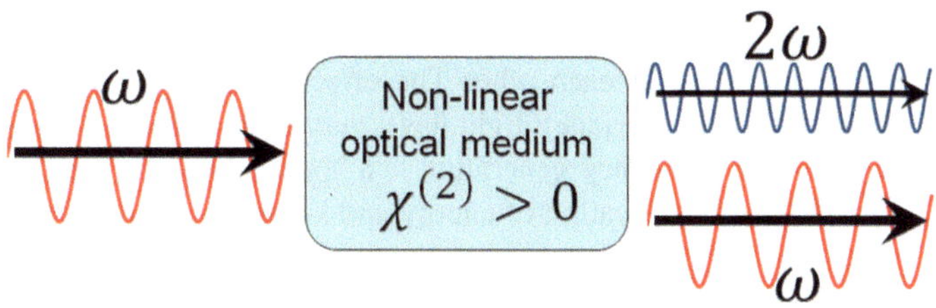

Fig. 2.2 Schematic of the SHG conversion of an excited wave in a non-linear medium

can be used as a spectroscopy technique to detect molecular monolayers adsorbed on surfaces [64]. Heinz and Shen adsorbed the single-layer laser dye rhodamine planar fused silica surface; then the surface was coated by nanosecond ultrafast laser. The SHG light having the characteristic spectrum of the adsorbed molecules and its electronic transition was measured as reflection from the surface, and the quadratic power dependence on the pump laser power was demonstrated.

In biology and medical science, the effects of second harmonic generation are used in high resolution optical microscopy. Due to the non-zero second harmonic coefficient, only non-central symmetrical structures are capable of emitting SHG light. One such structure is collagen, which is found in most load bearing tissues. Using a short pulse laser such as a femtosecond laser and a suitable set of filters, the excitation light can be easily separated from the transmitted frequency doubled SHG signal. This allows for very high axial and lateral resolutions, comparable to confocal microscopes, without the need for pinholes. SHG microscopy has been used to study the cornea and the sclera, which are mainly composed of collagen [65]. The generation of the second harmonic can be produced by several non-centrosymmetric organic dyes; however, most organic dyes also produce incidental fluorescence as well as second harmonic generation signals [66]. Up to now, it has been shown that only two classes of organic dyes do not produce any collateral fluorescence and work purely on the second harmonic generation. Recently, using a two-photon excitation fluorescence and a second harmonic generation microscope, a group of Oxford University researchers have shown that organic porphyrin-type molecules can have two-photon fluorescence and second harmonics to produce different transition dipole moments [67]. It happens from the same transition dipole moment. Second harmonic generation microscopy is also used in materials science, for example to characterize nanostructured materials [68].

2.3.2 Sum-Frequency Generation (SFG)

The sum frequency generation (SFG) is a second-order nonlinear optical process based on two input photon quenching at angular frequency ω_1 and ω_2 At the same time, a photon is generated on the frequency ω_3. Like any second order $\chi^{(2)}$ phenomena in nonlinear optics, which can only occur under the following conditions: light and matter interact with each other The effect is that the material is asymmetrical (e.g., surface and interface); light has a very high intensity (usually from a pulsed laser). The sum frequency generation is a "parametric process", means that the photon satisfies the conservation of energy and keeps the material unchanged:

$$\hbar\omega_3 = \hbar\omega_1 + \hbar\omega_2 \tag{2.3.2}$$

SHG is a special case of SFG, and its polarizability can be written as:

$$\chi^{(2)} = \beta(-(\omega_1 + \omega_2); \omega_1, \omega_2) \tag{2.3.3}$$

2.3.3 Raman Amplification

Raman amplification is based on the stimulated Raman scattering (SRS) phenomenon, when a lower frequency "signal" photons induce inelastic scattering of higher frequency "pump" photons in a nonlinear system in optical media. As a result, another "signal" photon is generated, and the remaining energy is resonantly transmitted to the vibration state of the medium. As with other stimulated emission processes, this process allows for full light amplification. At present, optical fiber is mainly used as a SRS nonlinear medium for telecommunications; in this case, it is characterized by a resonance frequency downshift of 11 THz (corresponding to a wavelength shift of 90 nm at 1550 nm). The SRS amplification process can be easily cascaded, thus essentially accessing any wavelength in the fiber low loss pilot window (1310 and 1550). In addition to applications in nonlinear and ultrafast optics, Raman amplification is also used for optical communications, allowing full-band wavelength coverage and online distributed signal amplification. The SRS phenomenon is not only used in the characterization of chemistry, biomedicine and materials science, but also the main principle of many Raman lasers.

2.3.4 Four-Wave Mixing

Four-Wave Mixing (FWM) is an intermodulation phenomenon in nonlinear optics in which the interaction between two or three wavelengths produces two or a new wavelength. It is similar to the third-order intercept point in electrical systems. Four-wave mixing can be compared to intermodulation distortion in standard electrical systems. This is a parametric nonlinear process because the energy of the incident photons is conserved. FWM is a phase sensitive process because the efficiency of the process is strongly influenced by the phase matching conditions.

Figure 2.3 is an energy level diagram of a non-degenerate four-wave mixing process. The highest energy level can be true atomic or molecular level (resonant four-

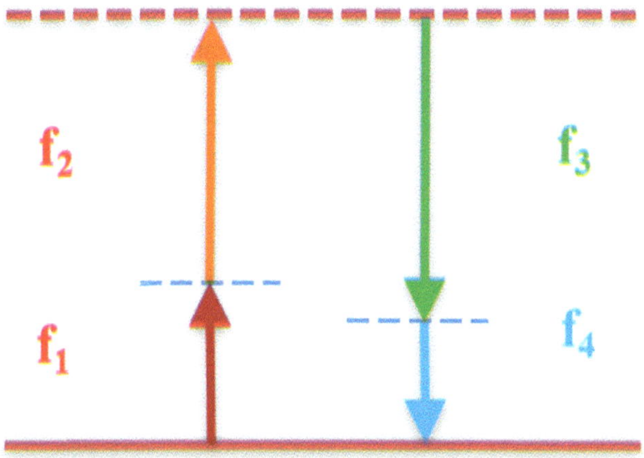

Fig. 2.3 Energy level diagram for a non-degenerate four-wave mixing process

wave mixing) or virtual level, far detuned non-resonant. The figure depicts a four-wave mixing interaction between frequencies ω_1, ω_2, ω_3 and ω_4.

Two common forms of four-wave mixing are known as sum frequency generation and difference frequency generation. In the sum frequency generation, three fields are input, and the output is a new high frequency field of the sum of the three input frequencies. In difference frequency generation, the typical output is the sum of two minus three.

The condition for effectively generating FWM is phase matching: when they are plane waves, the correlation k vectors of the four components must be zeroed. This becomes significant because the sum frequency and difference frequency generation are often enhanced when utilizing resonance in the mixed medium. In many configurations, the sum of the first two photons will be tuned to near resonance. However, near resonance, the refractive index changes rapidly and the addition of four collinear k-vectors cannot be accurately added to zero—so long mixed path lengths are not always possible because the four components lose phase lock. Therefore, the beam typically focuses both on intensity and focus on shortening the mixing area.

A frequently overlooked complexity in gaseous media is that the beam is rarely a plane wave but is usually focused on additional intensity, which adds an additive pi phase shift to each k vector under phase matching conditions. This requirement is often difficult to meet in a harmonic configuration, but is easier to satisfy in a differential frequency configuration where the pi phase shift is cancelled. As a result, the difference frequency is usually more widely adjustable and easier to set than the sum frequency generation, making it preferable as a light source even if its quantum efficiency is lower than the sum frequency.

Common stimulated Raman scattering (SRS) and coherent anti-Stokes Raman scattering (CARS) are FWM processes. The polarizability corresponding to the SRS can be expressed as:

$$\gamma^{(3)}_{SRS}\left(-\omega_{\mathrm{s}}, \omega_{\mathrm{p}}, -\omega_{\mathrm{p}}, \omega_{\mathrm{s}}\right) \tag{2.3.4}$$

And the CARS is:

$$\gamma^{(3)}_{CARS}\left(-\left(2\omega_{p} - \omega_{s}\right), \omega_{p}, \omega_{p}, -\omega_{s}\right) \tag{2.3.5}$$

Chapter 3
The Principle, Application and Imaging of CARS

3.1 Principles of CARS

Coherent anti-Stokes Raman spectroscopy, also known as coherent anti-Stokes Raman Scattering Spectroscopy (CARS), is a spectroscopy form primarily used in the analysis of chemical, physical, biomedical and related fields. It is sensitive to the same vibrational characteristics of the molecules seen in the Raman spectrum, usually the nuclear vibration of chemical bonds. Unlike Raman spectroscopy, CARS uses multiple photons to resolve molecular vibrations and produce coherent signals. Therefore, CARS is orders of magnitude stronger than spontaneous Raman emissions. CARS is a third-order nonlinear optical process beam involving three lasers: a pump beam p of frequency ω_{pump}, a beam of Stokes frequency ω_{stokes} and a probe beam at a frequency ω_{probe}. These beams interact with the sample and produce an anti-Stokes frequency $(\omega_{probe} + \omega_{pump} - \omega_{stokes})$ in a coherent optical signal. The latter is resonantly enhanced when the frequency difference between the pump and the Stokes beams $(\omega_{pump} - \omega_{stokes})$ coincides with the frequency of a Raman resonance, which is the basis of the technique's intrinsic vibrational contrast mechanism [69]. Advances in optical imaging technology have revolutionized our ability to study the microworld. Simple microscopy techniques, such as brightfield and differential interference contrast microscopy, play an important role in cell and molecular biology experiments, but do not provide chemical specificity. The ability to identify specific molecules in imaging has significantly improved our understanding of microscale biological processes. However, many of these techniques require the use of exogenous markers that often disrupt the system of interest. Intrinsic imaging techniques such as natural fluorescence imaging provide molecular specificity, but the number of endogenous fluorophores is limited. However, Raman microscopes do have significant limitations. The Raman effect is very weak); therefore, the data acquisition time is very long. Raman microscope images require high laser power and a long integration time of 100 ms to 1 s per pixel. These factors severely limit the application of Raman microscopes in life system research.

M. Sun et al., *Linear and Nonlinear Optical Spectroscopy and Microscopy*,
Progress in Optical Science and Photonics 29,
https://doi.org/10.1007/978-981-99-3637-3_3

3.1.1 Mechanism of CARS Signal Generation

In the CARS process, a pump beam of frequency ω_p and a Stokes beam of frequency ω_s interact with the sample through a wave mixing process. When the beat frequency $\omega_p - \omega_s$ matches the frequency of the Raman active molecular vibration, the resonant oscillator is coherently driven by the excitation field, producing a strong anti-Stokes signal at $\omega_{as} = 2\omega_p - \omega_s$ (Fig. 3.1). The Reintjes team at the Naval Research Laboratory is the first team to use CARS As a contrast mechanism for the microscope. Due to technical difficulties, the CARS microscope was not further developed until 1999 when the Pacific Northwest National Laboratory was revived with a new method. Since then, CARS mirroring has been used to visualize live cells in contrast to different vibration modes, including phosphate tensile vibration (DNA), amide I vibration (protein), and OH stretching vibration (water). And the tensile vibration (lipid) of the CH group. In these modes, the signal from the lipid is very high and a single phospholipid bilayer can be observed. At the same time, CARS has proven to be a powerful imaging modality for studying tissue in vivo. The advantages of CARS are summarized as follows:

1. It provides a comparison based on the inherent molecular vibration of the sample, avoiding the need for external labels.
2. It is more sensitive than a spontaneous Raman microscope, allowing for video vibration imaging and medium excitation power.
3. The nonlinear nature of the CARS process automatically confers its ability to slice three-dimensionally, which is critical for imaging thick tissue or cellular structures.
4. The anti-Stokes signal is blue-shifted from the pump and Stokes frequencies and is therefore easily detected in the presence of single-photon fluorescence.
5. When using near-infrared excitation wavelengths, CARS microscopes can observe depths close to 0.4 mm, allowing imaging in thick tissues.

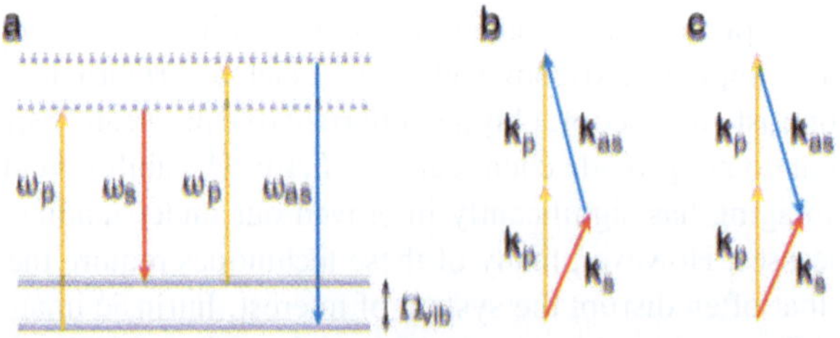

Fig. 3.1 **a** A diagram of the coherent anti-Stokes Raman scattering (CARS) process. When the difference between the pump and the Stokes frequency ω_{pump}, ω_{stokes} matches the molecular vibration frequency k_{vib}, the anti-Stokes signal is generated at the frequency $\omega_{as} = 2\omega_p - \omega_s$. **b** Phase matching conditions for the CARS generated in the forward direction. **c** Phase matching condition of CARS generated backward (epi-). k is called a wave vector and is given by $k = \frac{2\pi}{\lambda}$. Here, k_p, k_s, and cash represent pumps, Stokes, and anti-Stokes waves, respectively

6. Since the CARS process occurs in the terrestrial electronic state, sample optical simulation is minimized, especially when picosecond pulses are used to reduce multiphoton effects.

3.1.2 CARS Optical Configuration

Calculate the CARS signal detected by the forward $-(+z)$ and backward $-(-z)$ from equation. First, we consider a beam geometry that is commonly propagated. Figure 3.3a shows the CARS signal as a function of the diameter D of the spherical sample centered on the focus. The forward-detected signal first increases rapidly as the diameter increases, and then becomes saturated at diameters greater than 1.0 hp. The backward detection (or popularity) signal shows several interesting features. When the scatterer diameter D is much smaller than hp, it has the same amplitude as the forward signal. The first maximum is reached when the diameter D is equal to 0.3 hp. The increased oscillation behavior of the diameter is caused by the interference effect associated with the large wave vector mismatch in the backward direction. After the second maximum, the backward signal gradually decreases as the diameter increases. For scatterers with D = 8.0 hp, the backward signal is 105 times smaller than the corresponding forward signal. Thus, epidemiological geometry provides a means of imaging small features embedded in a nonlinear medium. 38, 39 is difficult to perform in the forward direction due to the presence of large forward signals in the surrounding solvent. A popular signal from scatterers, where y(3) is embedded in a nonlinear medium with $\chi^{(3)}$, exhibits the same behavior, but has an effective sample sensitivity of $\chi^{(3)}_{sca} - \chi^{(3)}_{sol}$. Signal generation from small scatterers provides a first contrast mechanism for popular CARS microscopes. Figure 3.3b shows the CARS signal for the hemisphere in the $z > 0$ region, centered at the focus. The forward detection signal shows the same behavior as the spherical sample. When D is equal to 0.5 hp, the maximum value of the popular signal appears. It can be seen that the popular signal from the boundary of such a semi-infinite sample is 1.2% of the forward detection signal. However, our calculations show that the CARS signal from the interface parallel to the optical axis advances and the radiated power is maximized along the optical axis. The signal generated at the interface perpendicular to the optical axis provides a second contrast mechanism for popular CARS microscopes. It should be mentioned that the popular CARS at the interface may also be caused by a mismatch in refractive index [$Re(\chi^{(3)})$]. Forward forward CARS on the index mismatched interface provides a third contrast mechanism for popular CARS images. In fact, if the excitation beam is not focused on the interface, the back-reflected signal is defocused on the detector and can be minimized by using confocal detection. For small scatterers, the retroreflected signal is negligible compared to the scattered signal from the scatterer. One way to avoid back-reflecting signals at the interface is to use a back-propagating beam geometry. We assume that the pump and Stokes beams propagate in the +z and −z directions, respectively. As shown in Fig. 3.3c, the changes in the forward (+z) and backward (−z) detection

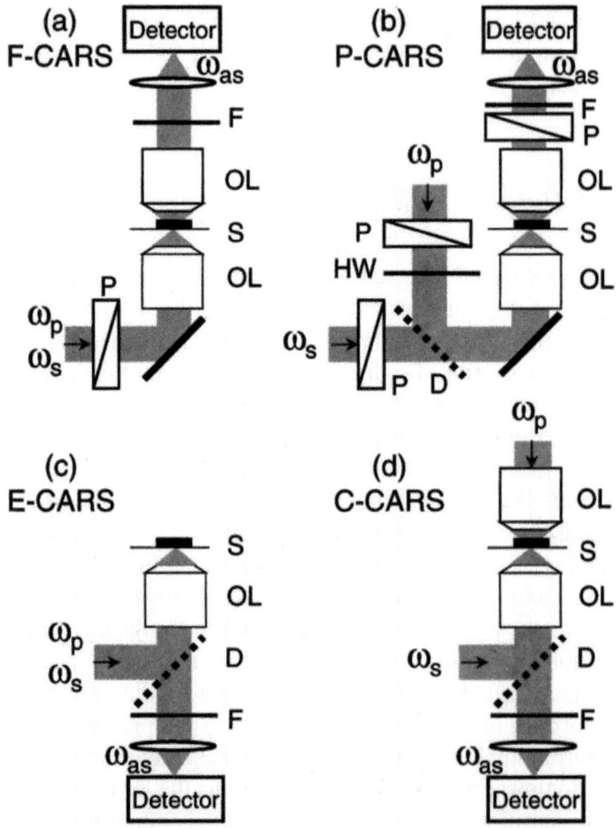

Fig. 3.2 Schematic diagram of the configuration of F-CARS, P-CARS, E-CARS and C-CARS microscopes. P, polarizer; OL, objective lens; S, sample; F, filter; HW, half-wave plate; D, dichroic mirror

signals and the sample diameter change in a manner similar to the popular signal in the co-propagating geometry, while the forward signal is much higher. In the backward signal. Furthermore, the maximum value of the forward detection signal in the backpropagation geometry is approximately twice that of the popular signal in the common propagation geometry. Backpropagation beam geometry provides another way to image small features and films embedded in nonlinear media by significantly reducing CARS signals from a large number of media. It should be noted that the CARS signal can also be generated at the interface of media having different $\chi^{(3)}$ values for backpropagation geometry (Fig. 3.2).

CARS microscopes were implemented in four different configurations: **a** forward detection CARS (F-CARS) with parallel polarization pump and Stokes beam, and **b** forward detection polarization CARS (P-CARS), **c** Popularity - CARS (E-CARS) is measured with parallel polarized pumps and Stokes beams, and **d** CARS (C-CARS) is backpropagated with parallel polarized pumps and Stokes beams and is in the

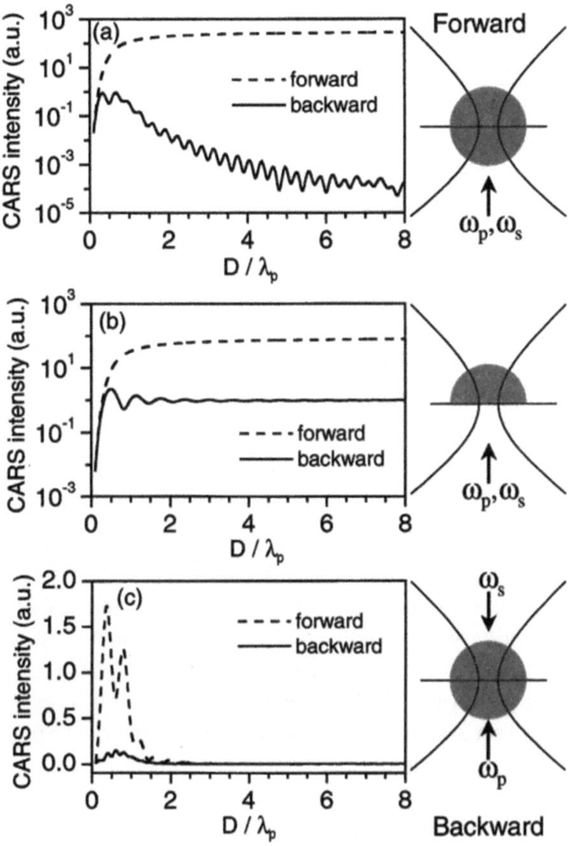

Fig. 3.3 **a** Forward and backward detection signals as a function of the diameter of the spherical sample in the geometry of the common propagating beam. **b** Same as in (**a**), but for hemispherical samples located in the z > 0 region. **c** The forward and backward detection signals are a function of the diameter of the spherical sample in the geometry of the backpropagating beam

pump The detection of the beam direction is performed. A schematic diagram of these configurations is shown in Fig. 3.2.

3.2 Biomedical Imaging of CARS

The development of the past few years has enabled CARS microscopes to be used in the fields of chemistry, materials, biology and medicine. Chemical applications include many studies on the sequencing of lipid vesicles, lipid layers and lipid domains. In the field of materials, CARS has been used to detect the kinetics of water in organic environments and has been applied to photoresist processing and liquid crystal sequencing. Recent exciting CARS applications have been in the field of biological and medical imaging and are the focus of this section.

While it is desirable to collect a complete spectrum for each object in a CARS microscope image, it is actually difficult to obtain these spectra. In a recent CARS microscopy experiment, a broadband femtosecond laser source was used in conjunction with a monochromator to collect pixel-by-pixel spectral data. For many samples, this results in an integration time of from milliseconds to seconds per pixel, resulting in significant photo damage. In addition, current detectors used in spectroscopy experiments have very long readout times, which limits their acquisition speed. These limitations currently limit the frame rate of 256 × 256 pixel images for a few minutes for CARS microspectral imaging experiments. For most biomedical experiments, such acquisition rates are too slow and are prohibitive for studying kinetics in biological systems. Even if the spectrum can be collected, it cannot be immediately converted to quantitative information without significant off-line processing.

Since the CARS intensity is proportional to $\chi^{(3)}$, the strength of the anti-Stokes signal can be written as:

$$I_{CARS}(\Delta) \propto \left|\chi_{NR}^{(3)}\right|^2 + \left|\chi_R^{(3)}(\Delta)\right|^2 + 2\chi_{NR}^{(3)} \operatorname{Re} \chi_R^{(3)}(\Delta) \quad (3.2.1)$$

where $Re[\chi^{(3)}]$ R is the real part of the resonance term of $\chi^{(3)}$. The first term has nothing to do with the Raman shift and is called a non-resonant background. The second term contains only resonance information and is a major contributor when detecting strong and/or concentrated resonant scatterers. The mixing between the non-resonant and resonance contributions produces a third term that contains the real part of the vibration response. The spectral response of each term is plotted in Fig. 3.4a, showing their respective contributions. Since the shape of the third term is dispersed, the addition of the three terms produces a red shift of the maximum of the CARS spectral peak and a negative fall at the blue end (25) (Fig. 3.4b). The red shift of the peak position depends on the relative intensities of the resonance and non-resonance contributions, so it is difficult to use a large amount of information in the Raman literature for specifying the CARS spectrum. The non-resonant contribution also introduces an offset that provides a background for the CARS microscope image (Fig. 3.4c, d). The blue end dip is not ideal because it gives a negative contrast (Fig. 3.4e). Spectral interference between two or more resonances may result in linear deformation and eliminates an immediate quantitative interpretation of the spectrum as adjacent peaks affect each other's intensity. In the crowded spectral region, this leads to a nearly unexplained CARS spectrum. Extraction of yttrium by interferometry can collect Raman spectra from CARS signals, although these methods may complicate the CARS imaging system.

CARS provides a new perspective on cell structure. A recent example is the imaging of plant cells. Plant cell walls are mainly composed of polysaccharides such as cellulose, lignin and glycoproteins. In the process of converting biomass into biofuels, lignin is primarily responsible for the chemical/enzymatic degradation of cellulose into short chain sugar molecules. However, it is difficult to image lignin using conventional imaging methods. In order to improve conversion efficiency, a chemical composition based contrast imaging technique is needed for real-time monitoring. The structure of lignin (Fig. 3.5a) produces a Raman spectrum (Fig. 3.5b)

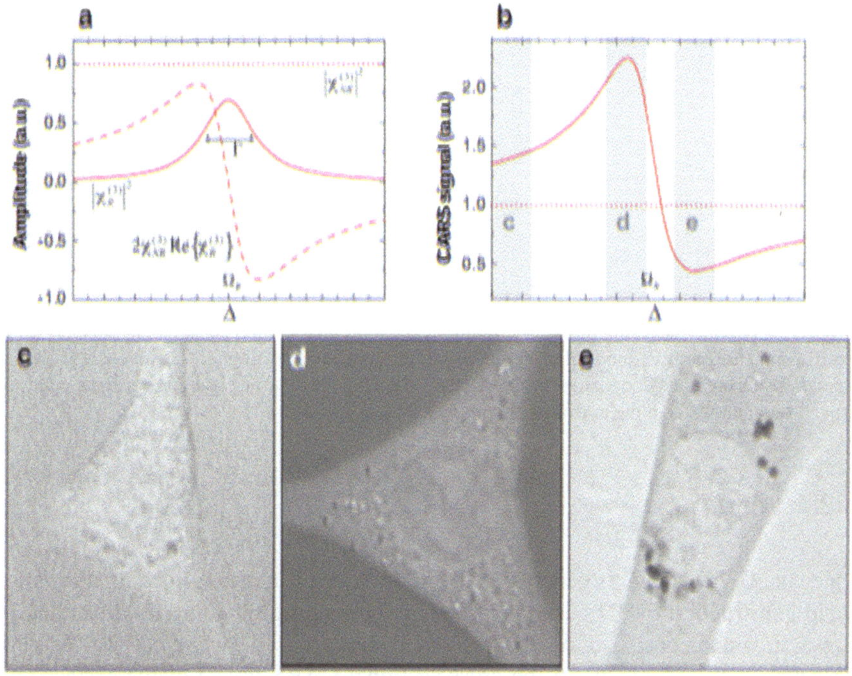

Fig. 3.4 **a** The three components of the coherent anti-Stokes Raman scattering (CARS) signal are plotted as a function of detuning. Shown here are pure resonance terms (solid lines), non-resonant background terms (dashed lines) and mixed terms (with discrete shapes) (dashed lines). The plotted curve is calculated as $\chi^{(3)}$. **b** Total CARS signal. A solid line represents the sum of the contributions of panel a, while a dashed line represents a non-resonant background. **c–d** Forward propagation CARS image of 3T3-L1 cells showing the contrast corresponding to the Raman shift region highlighted in figure b. Panel c shows the cells imaged by resonance; only non-resonant contrast was observed. Panel d shows a cell imaging and 2845 cm^{-1}, CH_2 symmetric stretching vibration. A variety of liposomes, including lipid droplets, are evident. **e** Cells imaged at 2950 cm^{-1} at the blue dip of the CH-stretch band. Resonance features appear darker on non-resonant backgrounds

which has a bandwidth of 1600 cm^{-1} due to the symmetrical vibration of the aryl symmetry ring, which can be used as a sensitive probe for lignin. Figure 3.5c shows a CARS image of corn stover adjusted to 1600 cm^{-1} stretch, revealing the distribution of lignin in a single cell wall.

In the past few years, many applications of CARS microscopes in biomedicine have emerged. CARS imaging is particularly useful for in vivo and in situ studies where the use of selectable markers may be impossible or prohibited. Compared to techniques such as magnetic resonance imaging, CARS has a small penetration depth; instead, it provides subcellular spatial resolution and high temporal resolution.

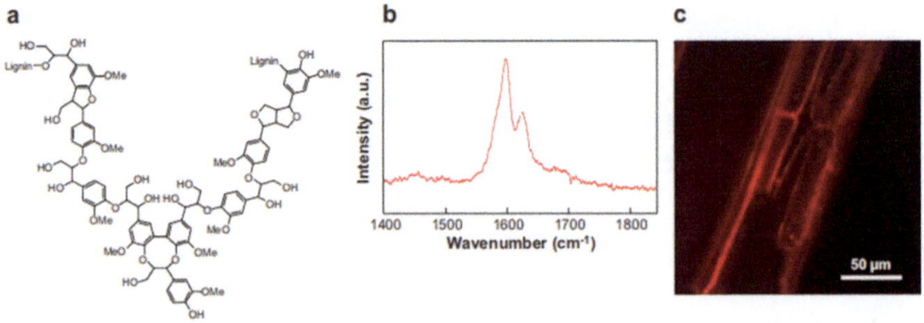

Fig. 3.5 **a** The chemical structure of the lignin polymer. **b** The Raman spectrum of has a remarkable band around 1600 cm^{-1} due to the stretching vibration of the aryl ring. **c** A coherent anti-Stokes Raman scattering microscope image at 1600 cm^{-1} showing the distribution of lignin in the cell wall around the plant cells in the corn stover

3.2.1 Lipid

Since the CARS signal is generated only at the focus, video rate imaging allows for rapid construction of 3D tissue maps. This 3D imaging capability is illustrated in Fig. 3.7f, which consists of 60 depth resolved slices spaced 2 μm apart. The stratum corneum is clearly visible in the cross-sectional CARS image of the surface and deeper sebaceous glands and fat cells in the tissue. The cross-sectional image looks similar to a tomogram obtained by optical coherence tomography (OCT), but unlike OCT produced by chemical contrast and provides higher lateral spatial resolution. CARS microscopes can also be easily combined with in vivo two-photon fluorescence microscopy to provide additional information. Figure 3.6 shows combined CARS and two-photon fluorescence images of mouse skin taken at a depth of 20 μm. A certain percentage of red blood cells have been labeled with DiD to highlight the capillary network. The spatial interaction between the blood vessels and the sebaceous glands is evident when the capillaries branch and wrap around the tissue structure.

In the past few years, the ability to CARS imaging has been greatly enhanced through a combination of laser engineering and microscopy optimization. Although the forward CARS image took 30 min to collect in 1999, we have demonstrated the ability to collect images at a rate of 30 per second in the epitaxial direction, with sensitivity increasing by nearly five orders of magnitude. Apparent detection of the high sensitivity of the CARS microscope makes it possible to study the vibrational selectivity of tissue in vivo. CARS microscopes are ideal for studying lipid distribution in tissues. Lipids and fat are distributed unevenly throughout the tissue and are stored in selected cell environments and organs. This heterogeneity combined with the high sensitivity of the CARS microscope to the CH vibration mode makes CARS an ideal choice for real-time studies of lipid and fat distribution in the body. As detection sensitivity is further improved, we anticipate that real-time imaging of the vibrational contrast of proteins and DNA will be achievable. The Raman reaction of DNA and protein is much weaker than lipids and requires an increase in

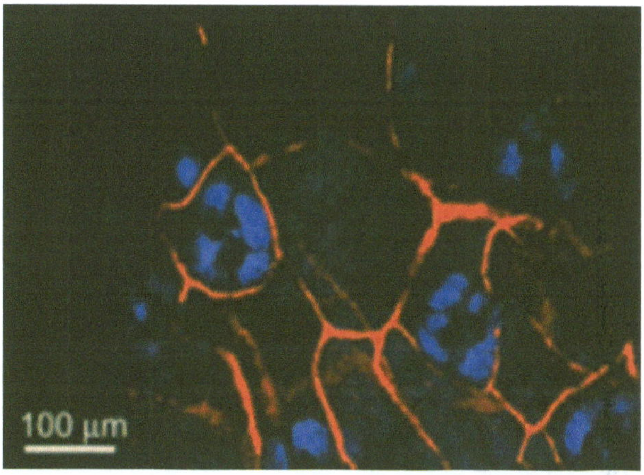

Fig. 3.6 Combine sequential CARS and two-photon fluorescence tissue images. The CARS signal is blue and the two-photon fluorescence is red. The Raman shift is set to 2,845 cm^{-1}, and the 816.7 nm pump beam drives the two-photon fluorescence excitation of the injected DiD dye. Sebaceous glands can be seen in the branch and annular capillary network

sensitivity of at least an order of magnitude. Emerging methods offer the prospect of increased sensitivity, taking CARS imaging to the next level. One of the main problems with tissue imaging is the ability to probe deep into the sample. The final penetration depth is limited by the working distance of the microscope objective. In this study, the maximum penetration depth of the mouse skin was found to be limited to a working distance of 125 μm, which is comparable to the depth achieved by two-photon excitation fluorescence using the same objective. The actual penetration depth of near-infrared radiation is likely to be affected by the deformation of the focus volume and the power reduction at the focus caused by the linear scattering of incident light in the turbid tissue. By using a long working distance objective, a penetration depth of a few millimeters has been achieved in a two-photon fluorescence microscope. We expect the CARS probe depth to increase when using such targets. In addition, penetration depth can be further increased by using longer wavelengths for CARS excitation to reduce the amount of scattering experienced in a turbid tissue environment. When collecting CARS images, we used an irradiation dose of 20 W/cm^2 (50 mW per bundle) on non-pigmented skin, which is much lower than 500 W/cm^2 of continuous wave radiation for pigmented skin. Damage threshold. Although nonlinear photodamage caused by high pulse peak power is a problem, we have not observed any signs of this effect in the tissue. We note that the peak power of the picosecond pulse sequence is lower than the femtosecond pulse sequence used in two-photon fluorescence tissue imaging.

In vivo CARS microscopy adds chemoselectivity to real-time non-invasive optical histology and may be useful for histopathology. In addition, epitaxial detection of CARS can be combined with fiber optic endoscopy for real-time guidance of molecular imaging and surgical intervention of intravascular atherosclerotic plaque.

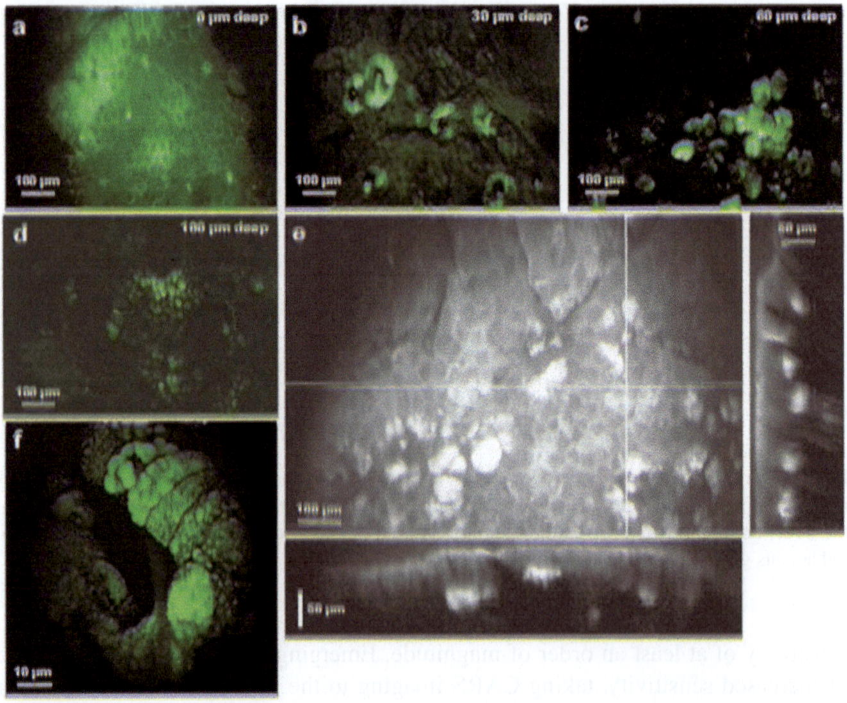

Fig. 3.7 Coherent anti-Stokes Raman scattering image of mouse skin at the lipid band (2845 cm^{-1}) in vivo. **a** Skin surface of hairless mice imaged on lipid strips. The outline of the keratinocytes is clearly visible due to the "mortar" of the lipid-rich cells in the stratum corneum. **b** Sebaceous glands are imaged at depths of 30μ. **c** Adipocytes of about 60 μm deep in the dermis. **d** Subcutaneous fat composed of many small fat cells, with a depth approaching 100μ. **e** A two-dimensional projection of 60 images of depth superposition is taken every 2 μm. The YZ and XZ cross sections (right and bottom plates, respectively) consist of a stack of depths along the white line. The cross section is rendered in reverse color to show better detail. **f** Three-dimensional rendering of mouse sebaceous glands. The crescent-shaped sebaceous glands surrounding the hair shaft are composed of a plurality of cells, each of which is filled with a plurality of micron-sized sebum-rich sebum particles

In vivo CARS imaging was first demonstrated on mouse skin and utilized a real-time video rate CARS imaging system. By adjusting the CH_2 vibration stretching frequency, the CARS microscope is able to visualize rich lipid structures throughout the 120μ depth of mouse ear skin. On the surface of the skin, a bright polygonal stratum corneum is visible due to the presence of intracellular "mortar" that binds many surface keratinocytes together. This intracellular material is rich in lipids, ceramides and cholesterol and produces a strong CARS signal (Fig. 3.7a). Multicellular sebaceous glands appear 20 μm below the surface of the skin (Fig. 3.7b). These glands are filled with micron-sized sebum particles, a compound rich in triglycerides and wax esters (Fig. 3.7e). At a depth of 60 μm, large fat cells are clearly visible and many are aligned along the blood vessels (Fig. 3.7c). At the bottom of the dermis, small fat cells forming a subcutaneous fat layer can be seen (Fig. 3.7d). The entire

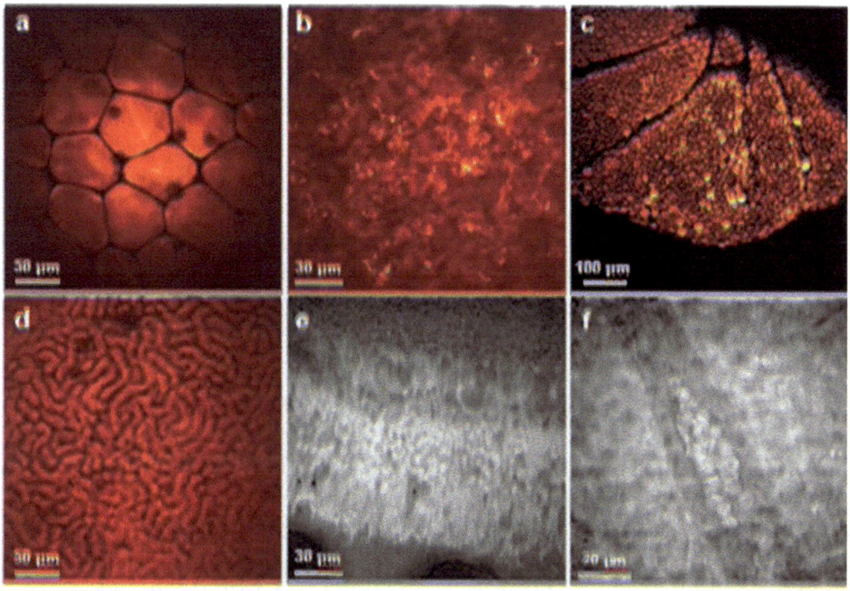

Fig. 3.8 Coherent anti-Stokes Raman scattering (CARS) imaging of various tissues in vitro compared to CH_2. **a** Epi-CARS images of mouse omental white adipose tissue. These large fat cells are filled with fatty acids and produce a strong CARS signal. **b** Epi-CARS microscopy of mouse lung tissue showing a single alveolar. The CARS signal is thought to come from lipid-rich surface active cells, Clara cells and macrophages. **c** Epi-CARS images of the kidney surface of mouse cells covered with adipocytes. **d** Epi-CARS images of mouse kidneys taken at a depth of 40 μm showed many renal tubules. **e** Cross section of the forward-propagating CARS image of the fixed bovine retina. The first few layers of the retina can be identified. **f** Epi-CARS images of fixed human retina photographed on the surface of the retina

tissue depth can be reconstructed quickly, three-dimensionally using the video rate CARS imaging system (Fig. 3.7f). The study was also able to spread into the skin by following the real-time tracking chemistry of baby oil.

The retina is composed of multiple layers of lipid-rich neurons, each with different functions and microstructures that can be easily identified using a CARS microscope (Fig. 3.8e, f). The photoreceptor, the inner and outer cores, and the inner and outer plexiform layers are easily visible in the cross-sectional image. The CARS depth stack allows for complete three-dimensional reconstruction of the retinal tissue, where the nerve fiber layer and ganglion cells can be visualized. Capillaries that pass through the surface of the retina, many containing red blood cells, are easy to see lipid contrast (Fig. 3.11f). Many CARS microscopy studies have focused on the structure and function of nerve bundles. For example, the resected spinal cord has been visualized using CH_2 stretching vibration , and the sciatic nerve of living mice has been imaged using minimal surgical techniques. Recent studies have even used CARS comparisons to study the destruction of neural structures in demyelinating diseases. A CARS microscope was also used to visualize the microstructure of the excised mouse lung (Fig. 3.8b). The lung tissue consists primarily of a small balloon called the alveoli, which is coated with a lipid-rich surfactant. CARS images of lung tissue adjusted to symmetric CH_2 stretching vibrations show that these alveoli and many lipid-rich cells, most likely surface-active cells (type 2 pneumonia), Clara cells and macrophages (66, 67) when using CARS When the microscope is imaged, the tissue of the kidney gives excellent contrast. Adipose tissue visualized on the surface of the kidney was prominent in the CARS image taken on the lipid strip (Fig. 3.8c). Below the surface of the kidney, at the depth of about 40 μm, the proximal and distal tubules are clearly visible (Fig. 3.8d). Careful examination of the small vessel wall revealed many round nuclei of the tubular epithelial cells that appear dark due to their low lipid content.

A new and exciting biomedical application of CARS microscopy is imaging of brain tissue. Brain tissue is lipid-dense because it consists of billions of neurons and supporting cells. Using CH_2 stretch contrast, CARS microscopy has been used to visualize many brain structures. The coronal section of the mouse at 2.8 mm from the anterior humerus showed many brain structures when imaged with a CARS microscope. To maintain cell resolution and image the entire organ, the brain mosaic shown was constructed from a CARS image of 700 μm × 700 μm (Fig. 3.9a). White matter bundles, such as the semi-oval center association network BER bundle, corpus callosum and corticospinal tract, are rich in myelin sheath and cause intense lipid CARS band signals. White matter regions in the diencephalon and deep brain nucleus can also be identified by their CARS signal intensity. To compare CARS brain samples from lipid-band CARS imaging to gold standard contrast (H&E) histological preparations of bioimaging, hematoxylin and eosin. Figure 3.9b shows a 700μ × 700μ corpus callosum and surrounding structure showing a comparison of gray matter compared to the corresponding H&E stained portion (Fig. 3.9c), revealing images of available information from the microscopic anatomy of the CARS microscope. The study also demonstrated that CARS can distinguish between healthy and disappearing brain tissue. Due to the lipid-poor nature of tumors, large astrocytomas are readily seen in lipid-band CARS images (Fig. 3.9d). A close examination of the edge of the tumor (Fig. 3.9e) reveals that astrocytoma is highly invasive because it can fuse the surrounding healthy white matter. These studies open the door to many potential clinical applications where CARS microscopy can one day replace traditional histopathology in brain imaging. In particular, the CARS microendoscope is capable of in-depth detection of brain tissue for diagnostic imaging and can reduce the need for brain tissue resection in the future.

CARS provides chemoselectivity. CARS can distinguish tissue structures based on their respective chemical composition. In the range of 2800–2900 cm^{-1}, the CARS spectra of sebaceous glands and deeper fat cells were almost the same (Fig. 3.10a). However, in the region of 2900–2970 cm^{-1}, the signal from the sebaceous gland is weak, while the strong signal from the adipocytes still exists. As a result, the CARS contrast of sebaceous glands and adipocytes at 2845 and 2956 cm^{-1} in Raman shift was significantly different (Fig. 3.10c–f).

The CH_2 vibration zone (2,800–3,100 cm^{-1}) consists of a number of vibrating bands, including aliphatic CH_2 and vinyl CH extensions. Therefore, the spectral

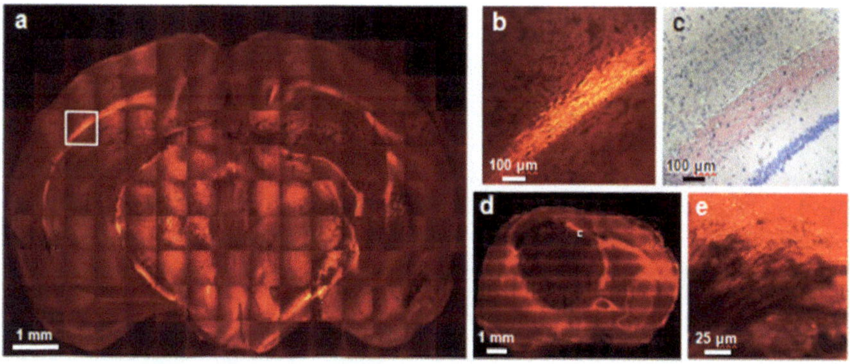

Fig. 3.9 Epi-CARS microscope is used for brain tissue imaging. **a** A mosaic image of a coronal section of a mouse brain taken at a lipid band showing many brain structures. **b** A single enlarged image corresponding to the white frame in panel a. **c** Hematoxylin and eosin (H&E) images of the same region of the same mouse brain. The structures visible in the two images, from top left to right, are the cortex, oriens layer and pyramid layer. The call body is a myelinated brain structure that produces a strong CARS signal. **d** A mosaic CARS image of astrocytoma in mice after four weeks of inoculation of tumor cells. **e** A magnified image corresponding to the white frame in panel d shows the filtered tumor at the edge

shape represents the chemical composition. Sebum consists mainly of triglycerides and other saturated fats and should have a large CARS signal at a symmetric CH_2 stretching frequency, approaching 2,845 cm^{-1}. Since fat cells contain higher concentrations of unsaturated fatty acids than sebaceous glands, changes in the CARS spectrum reflect differences in chemical composition.

To verify this result, mouse ears were sectioned to prepare 10 μm thin tissue sections, which can be probed in a Raman microscope. The isolated Raman spectra of various sebaceous glands and adipocytes support our distribution (Fig. 3.10b). In the 2870–2940 cm^{-1} of the Raman spectrum, the increase in the spectral intensity of the fat cells relative to the sebaceous glands leads to a shift in the high frequency of the CARS spectrum to a higher energy, which is the difference in CARS contrast. Therefore, even within the range of the CH stretching region, CARS can selectively highlight different chemical compositions. CARS microscopes track chemical diffusion in the skin. The unique imaging properties of the CARS microscope enable it to track the delivery of specific chemicals in the body's tissues in real time. To demonstrate this potential, we infiltrated externally applied mineral oil through the surface of mouse skin. The change in oil distribution can be observed by adjusting the CH_2 stretching vibration of the oil.

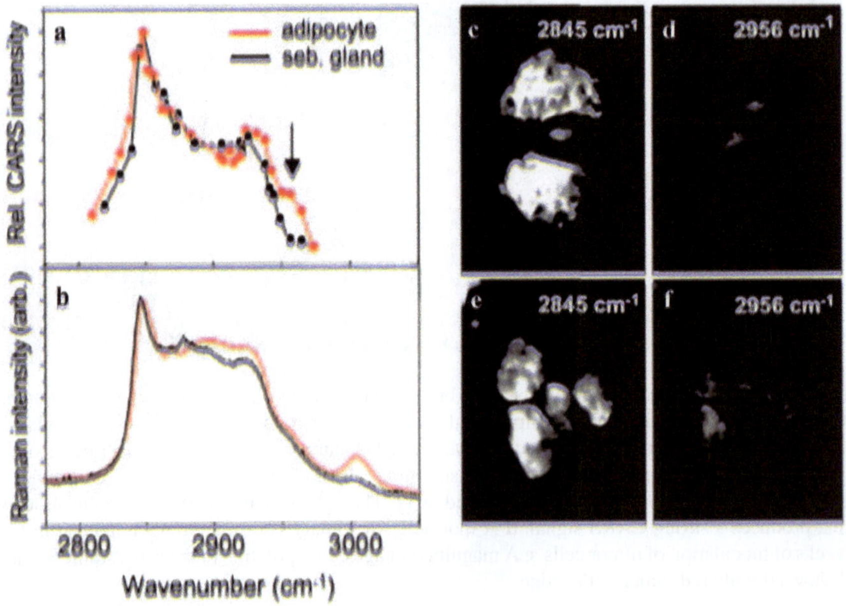

Fig. 3.10 Spectral differences between sebaceous glands and dermal adipocytes. **a** In vivo CARS spectra of sebaceous glands (black) and adipocytes (red) obtained by point-by-point wavelength scanning of the pump beam. Note the dissimilarity of the spectral intensities of higher wave numbers (2956 cm^{-1}, indicated by arrows) caused by different chemical lipid compositions. **b** Ex vivo Raman spectra of individual sebaceous glands (black) and adipocytes (red) recorded from 10 μm thick section tissue sections. We note that the CARS spectrum in the range of 2900–2970 cm^{-1} provides more spectral sensitivity to the saturation of the aliphatic chain than the spontaneous Raman spectrum. The CARS images of the (C and D) sebaceous glands were at 2845 cm^{-1} **c** and 2956 cm^{-1} (**d**). The CARS images of (**e** and **f**) fat cells were at 2845 cm^{-1} **e** and 2956 cm^{-1} (**f**)

3.3 Materials Imaging of CARS

CARS technology can also be used to characterize material properties. Since the internal lattice of the material also has special Raman signal peaks, CARS imaging technology can be used to characterize the surface morphology of the material. This approach has some advantages that traditional Raman imaging does not have. First of all, CARS technology uses nanosecond or even femtosecond lasers, so the surface topography can be characterized under the premise of ensuring that the material is not destroyed by the high-energy laser. Secondly, because the detected signal is on the anti-Stokes side, the background signal of some fluorescent materials can be fundamentally avoided. This provides a good condition for characterizing the surface of strong fluorescent materials.

3.3.1 CARS Image for Porous Carbon

Nanoporous carbon is an important carbon material in addition to diamond, fullerene, graphite, carbon nanotube and graphene. Although nanoporous carbon has great potential, some of its physical and chemical properties need to be adjusted in specific applications. Nitrogen doping is an effective strategy that can significantly change the properties of nanoporous carbon. N dopants usually include pyrrole-N, pyridine-N and graphite-N, and these N configurations can have a significant effect on carbon atoms. Generally, the doping of N into nanoporous carbon is based on post-processing or in-situ substitution. Recently, coordination chemistry has provided opportunities for the realization of clear nanoporous carbon. At present, solid-state pyrolysis is constructed by N-rich organic ligands. The coordination framework has been recognized as an effective method to achieve nanoporous carbon N doping. However, despite the great advantages of using coordination frameworks as precursors, controlling the N configuration in a predictable manner is still not impossible. Porous carbon materials not only have the advantages of high chemical stability and good conductivity of carbon materials, but also have the characteristics of high specific surface area, rich pore structure, and adjustable pore size due to the introduction of porous structure. They are used in catalysis, adsorption and electrochemical energy storage. All have been widely used. Li et al. [70] characterized porous carbon materials under a nonlinear optical microscope, see Fig. 3.11. Similar CARS images were obtained at 1587 and 1360 cm^{-1}. Both of the Raman shifts are characteristic peaks of carbon. Therefore the signal is concentrated on the porous carbon.

3.3.2 CARS Image for Graphene

Graphene is a two-dimensional carbon nanomaterial with a hexagonal honeycomb lattice composed of carbon atoms and sp^2 hybrid orbitals. Graphene has excellent optical, electrical, and mechanical properties, and has important application prospects in materials science, micro-nano processing, energy, biomedicine, and drug delivery. It is considered a revolutionary material in the future. The physicists Andre Gaim and Konstantin Novoselov of the University of Manchester in the United Kingdom successfully separated graphene from graphite using the micromechanical exfoliation method, and therefore jointly won the 2010 Nobel Prize in Physics. The common powder production methods of graphene are mechanical peeling method, redox method, SiC epitaxial growth method, and thin film production method is chemical vapor deposition (CVD). In fact, graphene exists in nature, but it is difficult to peel off a single layer structure. Layers of graphene are stacked to form graphite, and 1 mm thick graphite contains about 3 million layers of graphene. A lightly stroked pencil on the paper may leave a few layers or even just one layer of graphene. In 2004, two scientists, Andre Geim and Konstantin Novoselov, of the University of Manchester in the United Kingdom, discovered that they could use a very simple method to get

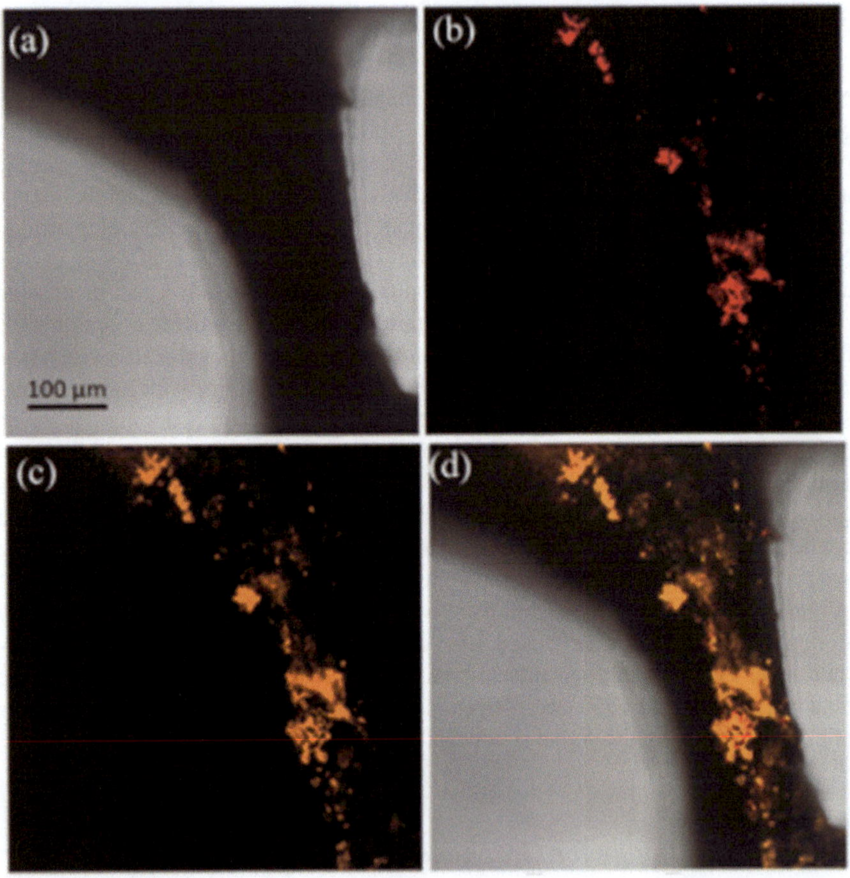

Fig. 3.11 **a** The bright field of porous carbon material, **b** and **c** CARS images at 1587 cm^{-1} and 1360 cm^{-1}, and **d** the merged images

thinner and thinner Graphite flakes. They peeled off the graphite flakes from the highly oriented pyrolytic graphite, then glued the two sides of the flakes to a special tape, and peeled off the tape to split the graphite flakes in two. Keep doing this, so the flakes become thinner and thinner, and finally, they get a flake composed of only one layer of carbon atoms, which is graphene. Since then, new methods of preparing graphene have emerged endlessly. In 2009, Andrei Geim and Konstantin Novoselov discovered the integer quantum Hall effect and the quantum Hall effect at room temperature in single-layer and double-layer graphene systems, respectively. Won the 2010 Nobel Prize in Physics. Before the discovery of graphene, most physicists believed that thermodynamic fluctuations did not allow any two-dimensional crystals to exist at a finite temperature. Therefore, its discovery immediately shocked the academic community of condensed matter physics. Although both theoretical and experimental circles believe that a perfect two-dimensional structure cannot exist

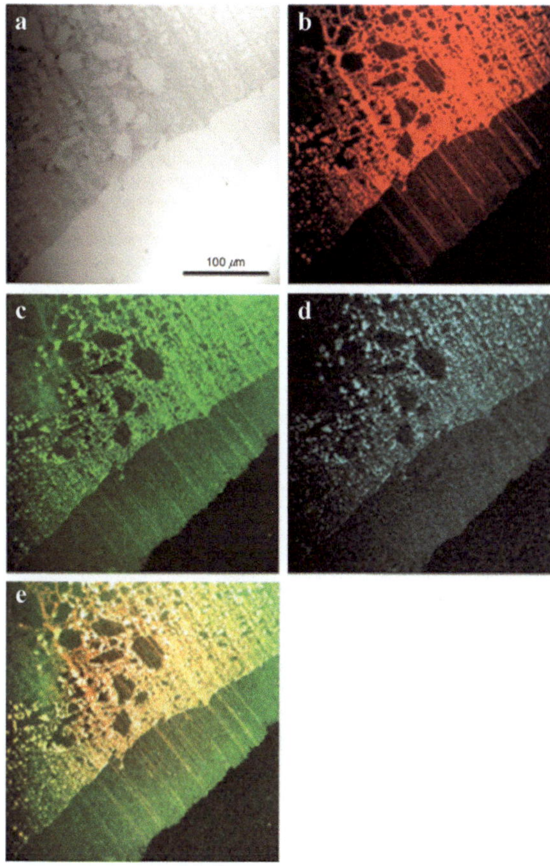

Fig. 3.12 CARS (red), TPEF (green) and SHG (blue) imaging of multilayer graphene

stably at non-absolute zero degrees, single-layer graphene can be prepared in experiments. On March 31, 2018, China's first fully automated mass-produced graphene organic solar optoelectronic device production line was launched in Heze, Shandong. The project mainly produces graphene organic solar cells that can generate electricity under low light (hereinafter referred to as graphene OPV). The three major solar power problems of application limitation, sensitivity to angle, and difficulty in modeling are solved.

In 2018, Li et al. [71] obtained CARS, TPEF and SHG imaging of multilayer graphene at the same time, see Fig. 3.12. It can be found from the figure that the nonlinear optical microscope can well characterize the interlayer relationship between multilayer graphene. The yellow area in Fig. 3.12e is the superposition of green and red, so it shows that the defect holes in graphene are the main areas that generate fluorescence and CARS signals.

Chapter 4
The Principle, Application and Imaging of SRS

4.1 Principles of SRS

Stimulated Raman spectroscopy, also known as stimulated Raman scattering (SRS) is a spectroscopy technique used in physics, chemistry, biology and other fields. The generation mechanism is similar to that of spontaneous Raman spectroscopy: an excited photon with a corner frequency of ω_p, when absorbed by the molecule, has a certain probability to induce a vibration (or rotation) transition (Unlike inducing a simple Rayleigh transition). This will result in a photon with an offset frequency emitted by the molecule. However, SRS, unlike spontaneous Raman spectroscopy, is a third-order nonlinear phenomenon that requires the second Stokes photon (angular frequency ω_S) to stimulate the transition of a specific frequency. When the frequency difference between two photons ($\omega_p - \omega_S$) is equal to a specific Vibration (or rotation) transition (ω_ν), the number of such transitions will increase resonantly. In SRS, changes in excitation and Stokes light intensity can be regarded as signals. Selecting a constant frequency laser as the excitation light and scanning frequency laser as Stoke light (or vice versa), the spectral characteristics of the molecule can be obtained. This spectral feature is different from that obtained by other spectral methods: such as Rayleigh scattering. Because the exclusion rules applicable to the Raman transition are different from those applicable to the Rayleigh transition.The principle of SRS can be intuitively understood by the energy levels of molecules. Initially, the molecule is in the ground state (the lowest energy level state), then it absorbs both the excitation photon and the Stokes photon, and then a certain probability of vibration (or movement) transition occurs as a result. This transition can be regarded as a two-step transition. The first step is that the molecule is excited by an excited photon to a virtual state, and the second step is that the molecule releases to a vibration (or rotation) state close to the ground state. The virtual state, which is actually the end of the superposition of the real state, cannot be occupied by molecules. However, the simultaneous absorption of two photons may provide a way to connect the initial and final states, allowing the molecule to appear to be in an intermediate virtual

M. Sun et al., *Linear and Nonlinear Optical Spectroscopy and Microscopy*,
Progress in Optical Science and Photonics 29,
https://doi.org/10.1007/978-981-99-3637-3_4

state. When the energy difference between excited and Stokes photons is exactly equal to the energy difference between some vibrational (or rotational) states and the ground state, the transition probability through this excited process will increase by several orders of magnitude. The stimulated Raman scattering spectroscopy imaging technology is an organic combination of Raman scattering spectroscopy technology, stimulated emission technology and laser scanning confocal microscopy imaging technology. Its signal intensity is linearly proportional to the concentration of the measured object, so stimulated Raman Scattering microscopy imaging can reflect the concentration of chemical substances. Figure 4.1b describes the stimulated Raman loss detection mode, that is, the physical mechanism of detecting the stimulated Raman loss; Fig. 4.1c is a schematic diagram of the system for implementing stimulated Raman scattering microscopy in a general laboratory. First, the pump light excites the electron to a virtual energy state, and the Stokes light induces the electron in the high energy state to return to the vibrational energy level, and at the same time emit a photon with the same wavelength. The final result is that the intensity of the pump light is weakened (while the Stokes light intensity is enhanced), and the intensity of the stimulated Raman scattering signal can be analyzed according to this light intensity change. The frequency difference between the pump light and the Stokes light determines the Raman frequency to be detected. In order to obtain the intensity of the stimulated Raman scattering signal, a "modulation-demodulation" method is used to detect the reduction of Stokes light. The specific method is as follows: The Stokes light is modulated according to a specific frequency and then combined with the pump light, and then guided into the microscope with a scanning unit; when the Stokes light is "1", part of the pump light is converted into Stokes light, so the pump light itself will be weakened; when the Stokes light is "0", the pump light and Stokes light maintain their original intensity. In this way, the intensity of the pump light will have a specific frequency change. The two-dimensional stimulated Raman scattering microscopic image can be obtained by scanning with the galvanometer inside the microscope. After the pump light and Stokes light have stimulated Raman scattering at the sample, the stimulated Raman loss produced The signal passes through the filter to filter out the Stokes light component, then enters the photodiode to form a photocurrent, and then is transmitted to the lock-in amplifier for demodulation. After the lock-in amplifier demodulates the stimulated Raman loss, the stimulated Raman scattering signal can be obtained, and then it can be imaged. Under normal circumstances, in the laboratory microscope, the second harmonic (SHG) and two-photon fluorescence (TPEF) imaging can be performed simultaneously through the photomultiplier tube, which can realize the multi-modal imaging function.

Figure 4.2 shows the spontaneous Raman spectrum, stimulated Raman loss spectrum, and coherent anti-Stokes Raman scattering spectrum (CARS) of retinol in alcohol. The stimulated Raman spectrum can be obtained by detecting the stimulated Raman scattering loss. spectrum. The stimulated Raman scattering spectrum has a strong similarity with the spontaneous Raman spectrum, and due to the existence of the non-resonant background signal, the coherent anti-Stokes Raman scattering spectrum has a distortion relative to the spontaneous Raman spectrum, which cannot be accurately reflected. Spectral information of spontaneous Raman spectroscopy.

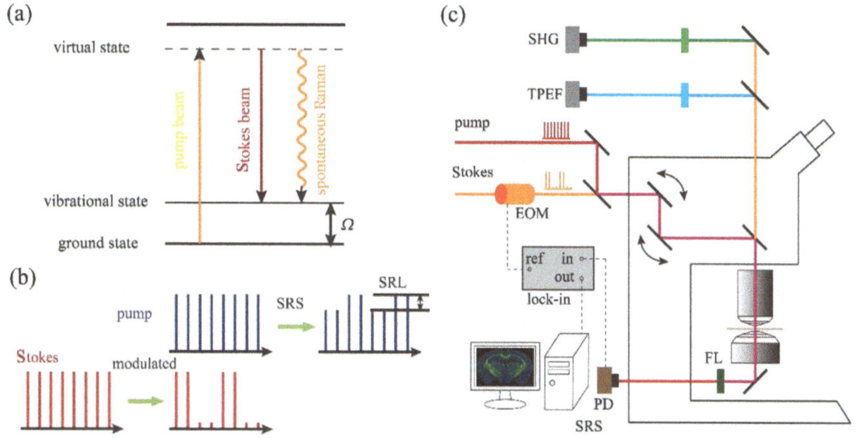

Fig. 4.1 A schematic diagram of the principle and implementation scheme of simple stimulated Raman scattering microscopy. **a** Schematic diagram of stimulated Raman scattering and spontaneous Raman scattering energy levels; **b** The detection mechanism of stimulated Raman loss; **c** Schematic diagram of stimulated Raman scattering experimental system [72]

Fig. 4.2 Femtosecond stimulated Raman scattering from a moving wave packet

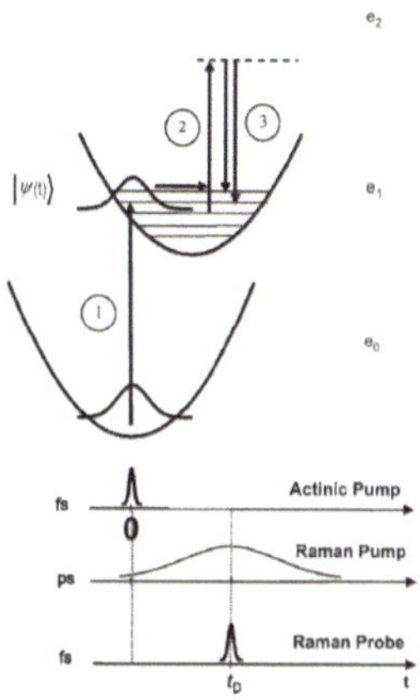

The difference between stimulated Raman scattering and spontaneous Raman scattering is that it requires two lasers (pump light and Stokes light) to act on the sample at the same time, and it can only target a certain Raman peak in the spontaneous Raman spectrum. For detection, the signal strength has been greatly improved (10^3 to 10^5 times) due to the existence of the stimulated emission process.

4.1.1 Quantum Theory of SRS

As shown in Fig. 4.2, the molecule is initially in the $\nu = 0$ state on the ground states surface e_0. An femtosecond actinic pump pulse (1) comes along and prepares the molecule as a moving wave packet on the excited state surface e_1. A picosecond Raman pump pulse (2) coupled with a femtosecond probe pulse (3) interrogates the moving wave packet, mediated by a higher excited state e_2, at carious times t_D, through stimulated Raman scattering as measured in the gain or loss of the probe spectrum. The excited processes (e_1 and e_2) on the manifolds of vibrational wave function. Therefore, the density matrix of this process is defined by:

$$\rho(t) = \sum_{a,b=1}^{2} |e_a\rangle\, \rho_{ab}(\mathbf{Q}, t)\, \langle e_b| \tag{4.1.1}$$

where the $\mathbf{Q}$ is the nuclear coordinates and $\rho_{ab}(\mathbf{Q}, t)$ is the vibrational density matrix associated with electronic states. The actinic pulse prepares a vibrational wave packet on electronic state e_1 is represented by:

$$\rho_{11}^{(0)}(t) = |\psi_1(\mathbf{Q}, t)\rangle\, \langle\psi_1(\mathbf{Q}, t)| \tag{4.1.2}$$

Then, the quantum Liouville equation of transition processes in SRS is given by:

$$\frac{\partial\rho}{\partial t} = -\frac{i}{\hbar}\left[\mathcal{H}_0 + \mathcal{V}(t), \rho(t)\right] - \Gamma\rho(t) \tag{4.1.3}$$

with

$$\begin{aligned}
&\mathcal{H}_0 = |e_1\rangle\, h_1(\mathbf{Q})\, \langle e_1| + |e_2\rangle\, h_2(\mathbf{Q})\, \langle e_2| \\
&\mathcal{V}(t) = -\boldsymbol{\mu}(\mathbf{Q}) \cdot \mathbf{E}(\mathbf{R}, t) \\
&\boldsymbol{\mu}(\mathbf{Q}) = |e_2\rangle\, \mu_{21}(\mathbf{Q})\, \langle e_1| + \text{c.c.} \\
&\mathbf{E}(\mathbf{R}, t) = \varepsilon_{pu} E_{pu}\,(t - t_D) \exp\left(i\mathbf{k}_{pu} \cdot \mathbf{R}\right) \\
&\quad +\varepsilon_{pr} E_{pr}\,(t - t_D) \exp\left(i\mathbf{k}_{pr} \cdot \mathbf{R}\right) \\
&= \mathbf{E}_{pu}\,(t - t_D) + \mathbf{E}_{pr}\,(t - t_D)
\end{aligned} \tag{4.1.4}$$

The third-order polarization is very important for the intensity of SRS, which is given with density matrix by:

$$P^{(3)}(t) = \mathrm{Tr}\left\{\mu_{21}^{*} \rho_{21}^{(3)}(t)\right\} + \text{c.c} \tag{4.1.5}$$

It should be noted the μ_{ab} is the coordinate-dependent transition dipole moment between electronic states e_a and e_b. The density matrix of perturbative response to third order is obtained using (4.1.3):

$$\rho_{21}^{(3)}(t) = \frac{i}{\hbar}\int_{t_a}^{t} dt' e^{-\gamma_{21}(t-t')} e^{-ih_2(t-t')/\hbar}$$
$$\times \left(\mu_{21}\cdot\mathbf{E}_{pu/pr}\left(t'-t_D\right)\rho_{11}^{(2)}\left(t'\right)-\rho_{22}^{(2)}\right.$$
$$\left.\times\left(t'\right)\mu_{21}\cdot\mathbf{E}_{pu/pr}\left(t'-t_D\right)\right)e^{ih_1(t-t')/\hbar} \tag{4.1.6}$$

where the t_a is a time after the actinic pump pulse is over ,but before the Raman pump and probe pulses appear. Therefore, the polarization of time-dependent SRS is defined by:

$$P_{\mathrm{RRS(I)}}^{(3)}(t;t_D) = e^{i\mathbf{k}_{pr}\mathbf{R}}\left(\frac{i}{\hbar}\right)^3\int_{t_a}^{t}dt_1\int_{t_a}^{t_1}dt_2\int_{t_a}^{t_2}dt_3 E_{pu}(t_1-t_D)E_{pr}(t_2-t_D)E_{pu}^{*}(t_3-t_D)$$
$$\times\langle\psi_1(t_a)|\,e^{(ih_1-\gamma_1/2)(t_3-t_a)/\hbar}\mu_{21}^{*}e^{(ih_2-\gamma_2/2)(t_2-t_3)/\hbar}\mu_{21}e^{(ih_1-\gamma_1/2)(t-t_2)/\hbar}$$
$$\times\mu_{21}^{*}e^{(-ih_2-\gamma_2/2)(t-t_1)/\hbar}\mu_{21}e^{(-ih_1-\gamma_1/2)(t_1-t_a)/\hbar}\,|\psi_1(t_a)\rangle+\text{c.c.} \tag{4.1.7}$$

4.1.2 Instrumentation of SRS

As shown in Fig. 4.3a, The SRG and SRL instrumentations are installed in a laser scanning microscope [72]. This configuration can not only obtain epi-SRG and epi-SRL imaging, but also can obtain epi-CARS images at the same time, see Fig. 4.3a. In the current device, a nearly linear concentration-SRL intensity relationship is obtained for retinol molecules, see Fig. 4.3b. By comparing with the spontaneous Raman spectrum, it can be found that the SRL spectrum has a very good agreement with the spontaneous Raman, see Fig. 4.3c. However, the CARS spectrum has a certain shift. This is because the coupling mode between the pump light and the Stokes light is different, and it is also related to the different nonlinear optical coefficients (see the Eqs. 2.3.4 and 2.3.5) of the corresponding wavelength.

4.2 Biomedical Imaging

4.2.1 SRS Imaging of Hela Cells

Figure 4.4a is the cell image at the non-Raman resonance frequency, and almost no signal can be seen in the figure; Fig. 4.4b is the cell image for the CH2 Raman vibration peak (2850 cm^{-1}), which mainly shows The distribution of lipids; Fig. 4.4c is a cell image of the CH3 Raman vibration peak (2928 cm^{-1}), which mainly shows the situation after the protein signal and the lipid signal are superimposed; Fig. 4.4d is the The distribution of the class (green) and the distribution of the protein (blue)

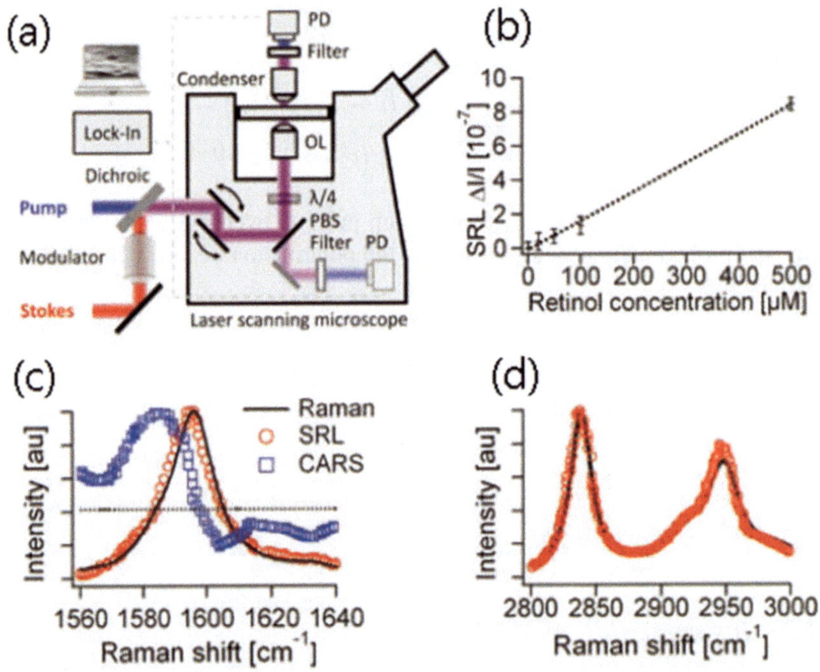

Fig. 4.3 **a** The label free stimulated Raman scattering microscopy. **b** The linear dependence of SRL and SRG on concentrations of retinol in ethanol at 1595 cm^{-1}. **c** The comparison of SRL, CARS and spontaneous Raman. **d** The comparison of SRL (red circles) and spontaneous Raman of methanol

are superimposed and false color is added. Compared with the spectrum in Fig. 4.4e, it can be seen that the spectrum of the nucleolus region is close to that of protein, and its composition is mainly protein, while the spectrum of lipid droplets may be lipid (oleic acid) spectrum and protein spectrum. The superposition, because it contains these two ingredients.

From the signal at 2800 cm^{-1} shown in Fig. 4.4a, it can be seen that during the stimulated Raman scattering imaging process, there is no interference of other non-resonant signals when Raman is out of resonance, and its spectrum is consistent with the corresponding spontaneous Raman spectra are basically the same, but there is a slight difference in spectral resolution. This is related to the laser light source used in the hyperspectral stimulated Raman scattering imaging system, and is not only determined by the properties of the stimulated Raman scattering process itself. Therefore, it is necessary to compare the spectrum in Fig. 4.4e with the stimulated Raman scattering spectrum and spontaneous Raman spectrum of a specific substance before the spectral shape of the measured object can be used to determine its chemical composition. In this way, the main components of the substances contained in the area can be judged according to the spectrum of the corresponding area in Fig. 4.4b and c. Therefore, using hyperspectral stimulated Raman scattering imaging technology, the

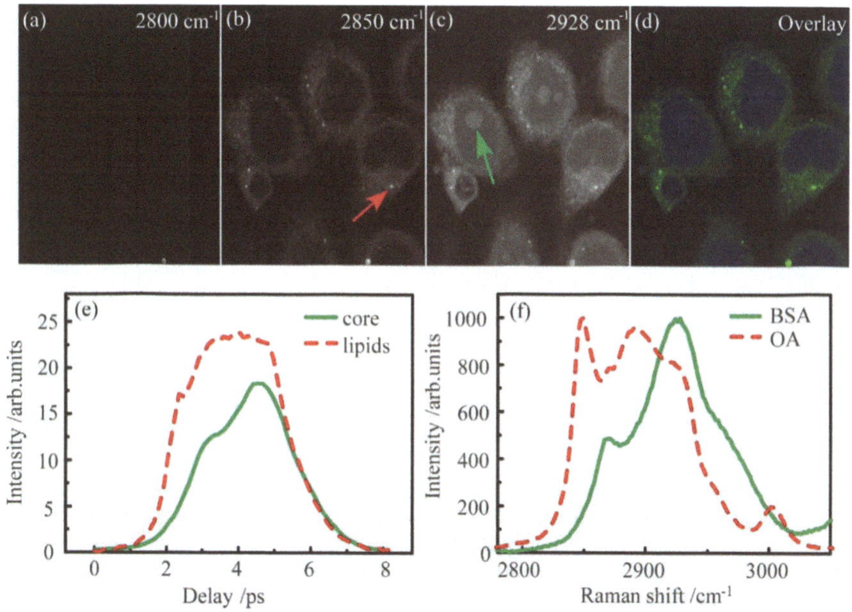

Fig. 4.4 Stimulated Raman scattering imaging and local area stimulated Raman scattering spectra of Hela cells at different Raman frequency shifts. **a–c** Stimulated Raman scattering images of cells at different Raman frequency shifts; **d** Protein and lipid synthesis images; **e** The lipid droplets and nucleolus regions in (**b**) and (**c**) Stimulated Raman scattering spectra; **f** Spontaneous Raman spectra of oleic acid OA and bovine serum albumin BSA [73].

corresponding Raman frequency shift can be realized by rapidly moving the optical delay line, thereby analyzing different components.

4.2.2 SRS Detection and Diagnosis of the Boundary of Glioma

Many progresses have been made in the imaging of brain tumors, and the imaging quality has been greatly improved. Evans et al. [74] used a monochromatic coherent anti-Stokes Raman scattering (CARS) microscope to detect the lipid content, and the results showed that: compared with normal brain tissue, the lipid concentration in the tumor area is much lower, and the signal The intensity is significantly reduced. However, this method of imaging only lipids lacks enough information for accurate diagnosis. Ji et al. [74] revealed the difference in Raman characteristics between normal brain tissue and tumor brain tissue, and found that dual-channel stimulated Raman scattering imaging based on lipid and protein content can provide key histo-

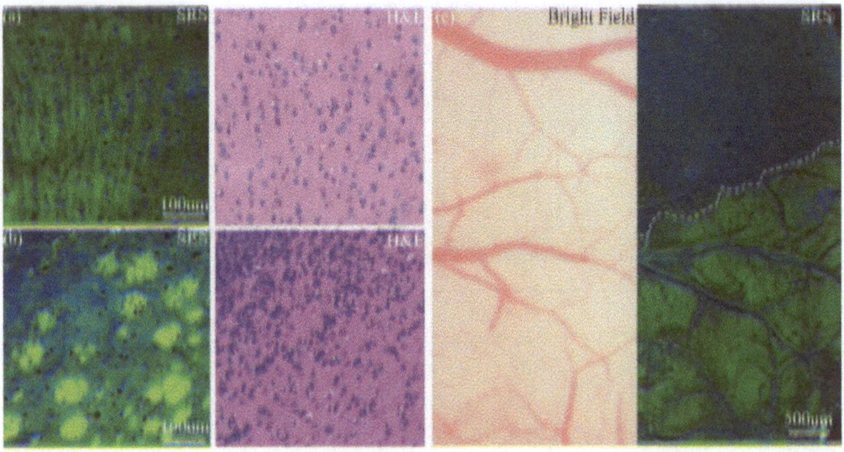

Fig. 4.5 Two-channel stimulated Raman scattering image of mouse brain xenotransplanted with human malignant glioma [34]. **a** Stimulated Raman scattering microscopic image and HE image of frozen section of normal brain; **b** Stimulated Raman scattering microscopic image and HE image of frozen section of malignant glioma penetrating brain; **c** Bright field What is seen in the microscope is the normal area, while the stimulated Raman scattering image clearly shows the boundary of the tumor [75]

logical information, including cell density And morphology, axon morphology, and the ratio of lipid content to protein content, etc.

Human glioblastoma ("GBM") xenografts have been used in early stimulated Raman scattering studies, and experiments have been carried out in vivo. The diagnosis of high-grade glioma depends on special histological features, including obvious nuclear heterogeneity, mitotic activity, cell necrosis and microvascular proliferation. Both frozen coronal brain tissue of normal mouse brain and human malignant glioma xenografts can be imaged by stimulated Raman scattering and HE. The ability of stimulated Raman scattering microscopy to identify tumor tissue is similar to the effect of HE staining, as shown in Fig. 4.5a, b [75].

4.2.3 SRS Imaging of Laryngeal Cancer

If the fresh tissue can be directly analyzed for histomorphology without labeling, the sample processing time can be reduced. Therefore, some researchers have performed stimulated Raman scattering microscopic imaging on fresh laryngeal cancer tissue that has not been frozen, sectioned, fixed, etc., and the morphology of fresh laryngeal cancer cells can be seen in the imaging results, as shown in Fig. 4.6. Shown as imaging of different areas of laryngeal cancer tissue, green represents lipids and blue represents proteins [76].

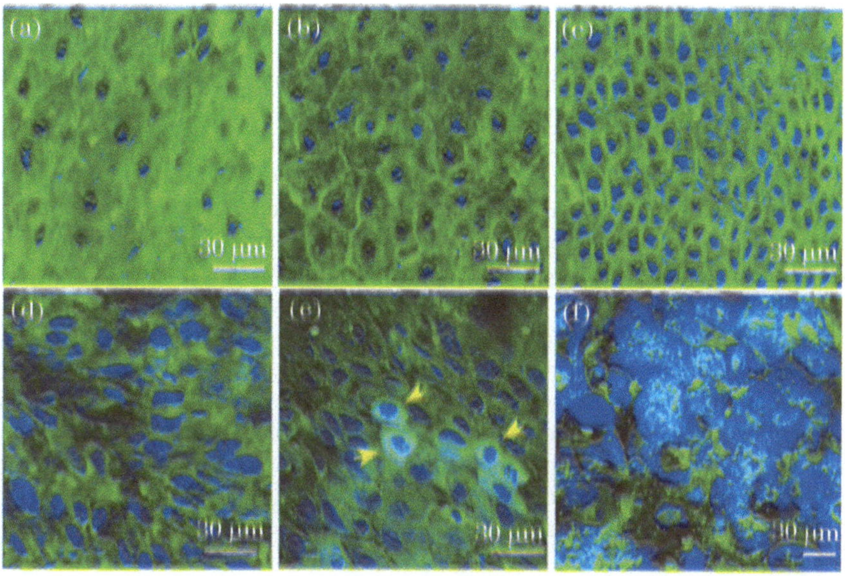

Fig. 4.6 Stimulated Raman scattering microscopic imaging of fresh larynx surgical tissue (lipid: green; protein: blue). **a–c** Normal squamous cells in different positions of the epithelial cell layer; **d** Enlarged nucleus and abnormal nucleus morphology [76]

From the stimulated Raman scattering microscopic imaging results shown in Fig. 4.6, it can be seen that the stimulated Raman scattering microscopic imaging can clearly distinguish the laryngeal cancer tissue, and it can be seen that different regions in the laryngeal cancer tissue have different Cell morphology and cell density have a good corresponding relationship with the gold standard HE in pathology.

4.3 Material Composition Analysis

Lithium metal batteries are next-generation energy storage batteries with huge application potential, because the theoretical specific capacity of lithium metal anodes is 10 times that of the current commercial lithium-ion battery graphite anodes. At the same time, it also has the lowest negative electrode potential among lithium battery materials, which makes it a perfect negative electrode material. However, due to its internal dendrite growth mechanism, lithium metal is also one of the most difficult materials to manipulate. This highly complex dendrite growth mechanism is still not fully understood by the scientific community, which may cause short circuit, fire or even explosion of lithium-ion batteries.

Although researchers know that the growth of lithium dendrites (needle-like lithium whiskers formed inside battery electrodes) is affected by the movement of ions in the electrolyte, they do not understand how ion transport and uneven ion concentration affect the deposition of lithium ions. The ion transmission imaging technology in the transparent electrolyte is very technically challenging, and the current technology cannot capture lithium ions in low ion concentrations.

Researchers from Columbia University today announced a research result [10]. They used Stimulated Raman Scattering Microscopy (SRS) for the first time to study the mechanism behind dendritic growth in lithium batteries. SRS is a technology widely used in biomedical research. This is the first time this technology has been applied to materials science research. By directly observing the ion migration in the electrolyte. They discovered a complete lithium deposition process. This process corresponds to three stages: no lithium ion consumption, partial consumption (previously unknown stage) and complete lithium ion consumption. At the same time, they also discovered a feedback mechanism between the growth of lithium dendrites and the heterogeneity of local ion concentration, so that the formation of lithium dendrites can be suppressed through the artificial solid electrolyte intermediate phase in the second and third deposition stages.

Studies have shown that there are three dynamic stages in the lithium ion deposition process:

1. When the ion concentration is much higher than 0, the slow and relatively uniform deposition of mossy Li; 2. Mixed growth of moss-like Li and dendrites; at this stage, the consumption of lithium ions occurs near the electrode, and lithium dendritic protrusions begin to appear; 3. Dendrite growth after complete exhaustion.

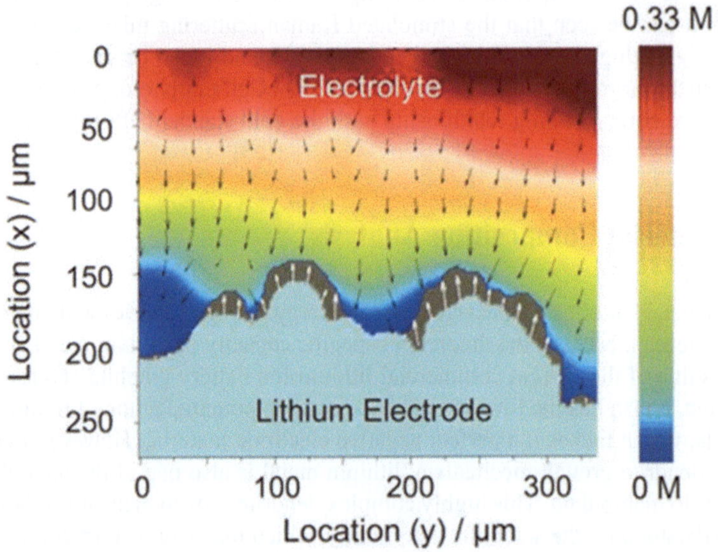

Fig. 4.7 SRS imaging shows the ion concentration distribution, ion flux and lithium deposition on the lithium surface [10]

When the surface lithium ions are completely exhausted, the lithium deposition will be dominated by "dendritic growth", and you will see the rapid formation of lithium dendrites.

Stage 2 is the key transition point at which the heterogeneous lithium ions on the lithium surface are depleted and induce the growth of lithium deposits from "moss-like lithium" to "dendritic lithium". At this stage, two regions begin to appear: the dendritic region where lithium begins to deposit dendrites at a faster rate, and the non-dendritic region where lithium deposition slows down or even stops (Fig. 4.7).

Chapter 5
The Principle, Application and Imaging of SHG

5.1 Principles of SHG

The second harmonic microscope (SHIM) is based on a nonlinear optical effect called second harmonic generation (SHG). SHIM has been established as a viable microscope imaging contrast mechanism for visualizing the structure and function of cells and tissues. The second harmonic microscope obtains the contrast from the change of the second harmonic coefficient of the incident light of the sample, while the traditional optical microscope obtains the contrast by detecting the change of the optical density, optical path or refractive index of the sample. SHG requires intense laser light to pass through materials with non-centrosymmetric molecular structures. The second harmonic light emitted by the SHG material is exactly half the wavelength (frequency doubled) of the light entering the material. Although two-photon excited fluorescence (TPEF) is also a two-photon process, TPEF loses some energy during the relaxation of the excited state, and there is no energy loss. Generally, inorganic crystals are used to generate SHG light, such as lithium niobate ($LiNbO_3$), potassium titanyl phosphate (KTP = $KTiOPO_4$) and lithium triborate (LBO = LiB_3O_5). Although SHG requires a certain material to have a specific molecular orientation in order to double the frequency of incident light. However, certain ordered large-scale non-centrosymmetrical biomaterials can still be highly polarized. Biomaterials such as collagen [70, 77], microtubules [78] and muscle myosin can generate SHG signals. The SHG mode is mainly determined by the phase matching condition. The common setting of the SHG imaging system is to use a laser scanning microscope with a Ti:Sapphire mode-locked laser as the excitation source. The SHG signal propagates in the forward direction. However, some experiments have shown that objects that produce approximately one-tenth of the wavelength of the SHG signal will produce almost equal forward and backward signals. Two-photon fluorescence (TPEF) is a completely different process from SHG: it involves exciting electrons to a higher energy level and then exciting them by photon emission (unlike SHG, although it is also a process involving two photons). Therefore, TPEF is an incoherent process, both

M. Sun et al., *Linear and Nonlinear Optical Spectroscopy and Microscopy*, Progress in Optical Science and Photonics 29,
https://doi.org/10.1007/978-981-99-3637-3_5

spatially (isotropic) and temporally (broad, sample-related spectra). Unlike SHG, it is also not structure-specific. Therefore, it can be coupled with SHG in multiphoton imaging to reveal certain molecules that can produce autofluorescence, such as elastin in tissues (for example, SHG can show collagen or myosin) [79]. Non-linear media that can generate second harmonics need to meet the following characteristics:

(1) No center of inversion symmetry. In the electric dipole approximation, the electric polarization intensity generated in the isotropic medium and the sample with the inversion symmetry center is equal and opposite, and they cancel each other, so the second harmonic will not be generated;

(2) It has a large second-order polarizability. The second harmonic intensity I_{SHG} is proportional to the square of the second-order polarizability $\chi^{(2)}$ of the nonlinear medium. Therefore, the medium must have a large second-order polarizability to be able to observe the second harmonic phenomenon. As mentioned in the previous article, the medium with the center of inversion symmetry has zero second-order polarization and will not generate second harmonics;

(3) Meet the phase matching conditions. In the case of phase matching, there is a strong mutual coupling between the fundamental wave and the harmonics, and the fundamental wave power can be effectively transformed into the second harmonic power, thereby ensuring a higher SHG generation efficiency.

In biomedical imaging research, the medium that meets the above requirements and can generate second harmonic signals is mainly divided into two categories: biomaterials and exogenous probes. Biological structural proteins usually assemble into large-area ordered and non-centrosymmetric structures, such as collagen, actin and tubulin. These structures are highly polarized and can generate SHG signals. In animals, the collagen family, as the main component of extracellular matrix, is the most abundant protein species. Fibrous collagen (Type I and Type II) is highly anisotropic. It is closely arranged by the repeating structure of collagen triple helix. The second harmonic generated by it provides relevant structural information for researchers. The second harmonic is polarization-sensitive. For many structural proteins of SHG signal sources, the amount of signal generated is related to the polarization state of the incident laser.

5.1.1 SHG Optical Configuration

Based on the above principles, researchers have introduced the nonlinear optical characteristics of the second harmonic into the biological imaging system. Compared with the common observable TPEF autofluorescence, SHG signal has the advantages of narrower spectrum, higher intensity and strict frequency doubling. The second harmonic microscope uses a laser in the near infrared region to excite the sample, and the long wavelength will produce a deeper penetration depth in biological tissues. Second harmonic imaging requires the use of femtosecond laser for two-photon excitation, which requires two photons to hit the sample molecule within femtosecond (10–15 s). The main advantage of this kind of excitation is the ability to limit the

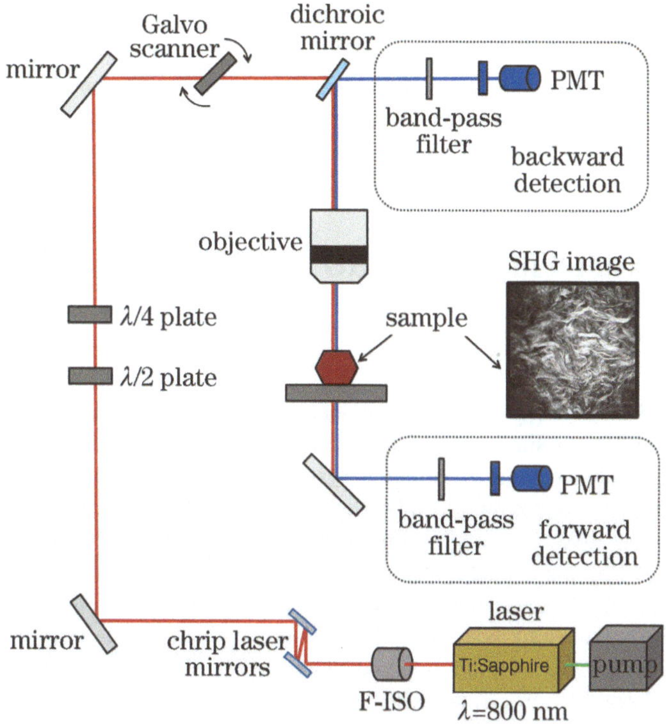

Fig. 5.1 Schematic diagram of SHG microscope device and image of mouse skin

excitation area to a tiny focal volume in a thick sample. The focal point of the objective lens is the only space with sufficiently high photon density, which can be used to improve the signal-to-noise ratio and spatial resolution. Fluorescence microscopes, such as confocal microscopes, only receive the emitted light at the focal point after being filtered by a pinhole, but single-photon excitation will cause the entire z-axis of the sample to emit light, resulting in greater phototoxicity.

The basic device of second harmonic imaging requires a scanning confocal microscope equipped with a femtosecond laser as the excitation light source. At present, there are many commercial multiphoton confocal microscopes that can be used for SHG imaging, and there are also self-assembled SHG imaging experimental devices [80]. These devices generally use a system similar to Fig. 5.1. The most commonly used laser is $Nd : YVO_4$ (532 nm; 5–18 W) pumped Ti: sapphire oscillator, the tuning range is about 700–1000 nm, the average power is 1–2 W, and the pulse width is about 100 fs. And through the $\frac{\lambda}{2}$ and $\frac{\lambda}{4}$ wave plates to adjust the phase of the incident light, see Fig. 5.1. SHG signal collection methods can be divided into two methods: forward detection and backward detection, and backward detection is commonly used. In order to avoid the influence of other bands of light and ensure that the collected signal is SHG signal, a filter must be used to filter out the interference signal. The polarization direction has an impact on the imaging results. Chen et al. [81] compared the SHG images obtained by linearly polarized light and circularly polarized light on the excitation of collagen fibers, and found that circularly polarized light can uniformly excite all directions, while linearly polarized light is more suitable for studying fibers.

5.2 Biomedical Imaging of SHG

5.2.1 Collagen

Multiphoton microscopy relies on nonlinear optical material interactions to provide contrast and optical slicing capabilities for high resolution imaging. Most multiphoton microscopy studies in biological systems rely on two-photon excited fluorescence (TPEF) to produce images. With the increasing use of multiphoton microscopy in thick tissue "live" imaging, second harmonic generation (SHG) from structural proteins has become a potentially important new contrast mechanism. However, SHG is typically detected in transmission mode, thus limiting TPEF/SHG registration and its practical application in thick tissue applications in vivo. In this study, we used a wide range of excitation wavelengths (730–880 nm) to demonstrate that TPEF/SHG registration can be easily achieved in unstained tissue by using a simple backscatter geometry. The combined TPEF/SHG technique is used to image the three-dimensional organ tissue model (RAFT). The structure and molecular origin of image-forming signals from various tissue components are determined by simultaneous spectral measurements and confirmation of immunofluorescence staining. Our results show that at shorter excitation wavelengths (<800 nm), the signal emitted from the extracellular matrix (ECM) is a combination of SHG and TPEF from collagen, whereas at longer excitation wavelengths, the ECM signal is completely attributed to SHG. The endogenous cellular signal is consistent with the TPEF spectra of the cofactors NAD(P)H and FAD at all excitation wavelengths. The reflected SHG intensity follows a quadratic dependence on the excitation power, exponentially decays with depth, and exhibits spectral dependence according to previous theoretical studies. The combined use of SHG and TPEF provides supplemental information that allows for non-invasive, spatially located in vivo characterization of cell-ECM interactions in unstained thick tissue.

The exact mechanism of salt ions and other charges The interaction of species and proteins is not fully understood. However, these interactions are known to regulate the solubility and aggregation processes of proteins. To date, much work has focused on the effects of ions on the water environment and describes how it affects the stability of proteins and the formation of nanostructures in solution. Few studies have explored the effects of interactions on self-assembled structures on a microscopic scale. An in situ non-invasive method is required to characterize the formed structure to understand the contribution of ion pair formation to the formation of higher order components of the microstructure within the bio-derived protein 3D scaffold.

Collagen is a major component of the extracellular matrix (ECM) in mammalian connective tissue. It accounts for about 20–30% of all proteins in the human body.1 Because collagen can form fibers that contribute to the stability of polymer networks, it is still a popular biomaterial choice for tissue engineering 2D and 3D scaffolds. Collagen undergoes complex molecular organization over a range of lengths, from triple-stranded helical collagen molecules to nano- (fibrils), micro- (fibers and fiber bundles) and macro= (gel) scales. Second harmonic generation (SHG) comparison as a valuable unlabeled spectral probe for non-invasive and in situ direct detection of aggregated collagen structures (fibers) within scaffolds. The SHG contrast is caused by the interaction between fibrous collagen and near-infrared (NIR) pulses, femtosecond laser scanning nonlinear microcopies. SHG is produced when photons interacting with fiber collagen combine to form new photons. Twice the energy. In addition to collagen filling within the material, the interaction between the laser pulse and the non-centrosymmetric triple helix structure of the collagen results in the scattering of tertiary (fibril) 4 and quaternary (fiber) 5 levels of collagen tissue, resulting in SHG contrast. Due to the use of near-infrared wavelengths, it is possible to generate and image this contrast deep within opaque 3D samples such as collagen hydrogels. A significant advantage of SHG contrast is the absence of bleaching, which has been used to successfully image structural proteins with high resolution and contrast in biomedical assessment of tissue structure.

We have used SHG to compare the contribution of different ions to collagen aggregation, which aggregates into a three-dimensional (3D) hydrogel to form a microscopic fiber structure. To the best of our knowledge, these studies are the first to attempt to directly, non-invasively and in situ systematically study the effects of various ions on collagen fibers and the hydrogel microstructure assembled with different salts. To gain insight into the structural changes associated with protein-added salts, we conducted such studies in a range of salt concentrations. We further combined our optical imaging experiments with conventional gel turbidity measurements to determine the time scale of the process and solubility involved to assess the amount of collagen remaining in the solution after fiber formation.

Second harmonic generation (SHG) imaging. The second harmonic generation image shows that the microstructure formed within the collagen hydrogel strongly depends on the concentration of phosphate present (Fig. 5.2). For example, when the phosphate concentration is 30 mM, the fibers formed are very small (about 1 mm, Fig. 5.2). When the phosphate concentration reaches 60 mM, there is little fiber length extension (up to 15 mm, Fig. 5.2). When further increasing the concentration of phosphate in the buffer solution, we observed a progressive trend in fiber length and width towards a larger value. For example, in 160 mM phosphate, the fiber length and width are about 40 mm and 18 mm, respectively Fig. 5.2). After further increasing the phosphate concentration, we observed a decrease in fiber width and length. For example, in 200 mM phosphate, the fiber length and width are about 17 mm and 1 mm, respectively.

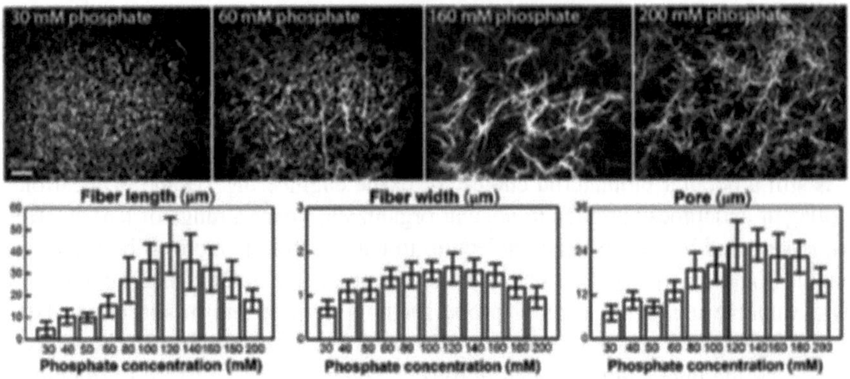

Fig. 5.2 (top row) Backscattered second harmonic generation (SHG) image of collagen hydrogel assembled from pH 7.4, 37 °C and different phosphate concentrations; (bottom row) quantification of collagen hydrogel parameters: fiber length, width And the effective aperture and the standard deviation corresponding to the average

Despite the many work of the past decade, the specific salting out effects of different ions on protein structures are poorly understood and lack systematic explanation. However, it is agreed that the type and concentration of added salt is the two processes that drive this interpretable behavior: (i) weakening electrostatic repulsion by specific association of ions with protein charged groups, and (ii) Change the tension of the ions. Protein-water-based interface. The correlation of anions with positively charged protein residues reduces the effective surface charge on the protein macromolecule. Larger (softer) anions are more effective at screening for electrostatic repulsion between protein molecules and promoting salting out behavior. In addition, the surface tension of water increases with the addition of salt , so the stable floc 18-collagen molecules and/or fibrils aggregate into a "fibrous" structure observed in the second harmonic generation (SHG) image. Thus, as the amount of salt in the final polymerization mixture increases, increased surface water tension and reduced electrostatic repulsion can potentially result in increased size of aggregated collagen floes (fibers).

At low phosphate concentrations (less than 20 mM), a stable collagen hydrogel was not formed and the sample was essentially liquid. Similar to previous studies, a stable gel could not be formed and we observed an increase in collagen solubility over this range of phosphate concentrations. Kuznetsova et al. mentioned that phosphate ions are strong fibril formation inhibitors in the range of 10–25 mM and lead to an increase in collagen solubility. Subsequent studies by Mertz et al. attribute this increase in solubility to the preferential interaction of the binary form of phosphate with collagen fibrils. The same study determined from infrared measurements that under physiological conditions, each collagen molecule in the fibril has one or two sulfate and dibasic phosphate binding sites, and monovalent phosphate does not. The researchers suggest that the bound divalent anion forms a salt bridge between the positively charged amino acid residues within the collagen fibers. However, this study further shows that this binding is insignificant for the formation, stability and structure of collagen fibrils and stabilizes the fibrils by preferentially excluding unbound anions in the interstitial water within the fibrils. It is proposed that NaCl can affect the binding of dibasic phosphates and the partitioning of dihydrogen phosphate in collagen fibers.

When the phosphate concentration is increased to about 30–60 mM, a stable hydrogel is formed. Similar to previous work, collagen assembly in diluted protein solution. When the phosphate concentration was increased by more than 30 mM, a decrease in the rate of collagen polymerization was observed. Fibers in the range of 30–60 mM phosphate concentrations are still small. The incorporation of larger fibrous structures within this range of phosphate concentrations is disadvantageous because only the hydrated shell of water surrounding the protein filled into the fibrous structure is moderately destroyed. When the phosphate concentration is high (80–160 mM), in addition to maintaining pH and binding to collagen molecules, phosphate must significantly compress the hydrated shell within the collagen fibers. The interaction of protein groups with salts may make fibrils more likely to stick to each other, forming larger "fibrous" flocs. This may explain why the fiber is larger at this phosphate concentration than the 30–60 mM phosphate induced sample. However, further increases in phosphate concentration (200 or 300 mM) again induced smaller 'fibrous' characteristics within the hydrogel. Very high phosphate concentrations (500 mM) result in the formation of fine precipitates and a slight increase in solubility. Proteins are often denatured by high concentrations of inorganic salts. Ions formed by the dissolution of inorganic salts may bind to the ionic groups of the protein and disrupt the interaction of the "fibrous" structure of the stabilized collagen. These interactions result in the formation of fine precipitates. Similar insights can be applied to sodium sulfate and sodium chloride-assisted collagen material assembly at 37 °C. In these samples, there is enough phosphate to maintain the pH and stabilize the collagen fibers. Similarly, a high concentration of 0.9 M NaCl appears to promote collagen solubilization and result in the formation of smaller "fiber" floes.

Typically in a self-assembling system, lowering or raising the temperature affects the formation of colloidal flocs in a different and inconspicuous manner. This further raises the need for direct, non-invasive and in situ measurements of these systems. The optical methods presented in this study are well suited for this task. In the assembly of sodium chloride-assisted collagen material at lower temperatures (27 °C and/or room temperature), we observed the formation of longer and thicker "fibrous" structures compared to the gel assembled at 37 °C. Lowering the temperature has the potential to improve the exothermic adsorption step, resulting in the formation of larger "fibrous" aggregates at 27 °C and room temperature. Interestingly, when a divalent anion such as sulfate is used, large and long fibers are formed at room temperature in a sulfate ranging from 5 to 20 mM. However, for these sulfate concentrations, a stable polymer network within the gel may not be established due to insufficient overlap between the formed fibers. Obtaining stable hydrogels in 75–100 mM sulphate is only challenging because they do not appear to polymerize even within a few days. When the sodium sulfate concentration is 150–300 mM, the total time required to polymerize the hydrogel decreases as the salt concentration increases. When the sulfate is from 5 to 20 mM, the fibers are significantly thinner and slightly longer.

The research presented in this study is timely. Recently, Li et al. also studied collagen molecular interactions, focusing on fibrils formed in various salts directly dissolved in a collagen solution adjusted to a suitable pH with 0.1 M NaOH. Based on low resolution transmission electron microscopy (TEM) images, the authors conclude that at pH 7.4, low concentration salts with monovalent ions result in the formation of filaments without a strip pattern. When the ionic strength was increased to 200 mM, the collagen fibrils were more ordered, and some large fibrils with a strip pattern were observed. When a salt having a multivalent ion is used, the collagen monomers aggregate to form an ordered bundle of fibrils, which exhibits a clearer strip pattern. A polarized light microscope image dispersed on a glass slide also obtained collagen fibrils formed in NaCl and KCl salts (100 mM, pH 7.4) and fibril bundles of about 80 mm in diameter detected in Na_2SO_4 and Na_2HPO_4 samples (ion strength) (250 mM, pH 7.4).

In a somewhat related study, Jiang et al. evaluated the effect of pH and electrolyte on the deposition of collagen into microcracks (approximately 3 nm anisotropic ribbon structure) deposited on newly cut mica supports. The deposition was controlled by using hydrodynamic flow, and the formed structure was examined by atomic force microscopy (AFM). The effect of $MgCl_2$, NaCl and KCl electrolytes on the assembly was investigated at a neutral pH of 7.5. Interestingly, low NaCl concentrations (50 mM) and high NaCl concentrations (200 mM) produced a single layer of collagen fibrils on the mica surface. However, when the NaCl concentration was 100 mM, the protrusions of the first layer merged into an incomplete second fibril layer.

Gobeaux et al. recently also studied the polymerization of collagen from concentrated solutions (40–300 mg/ml) at different pH and ionic strength. Two methods are used to adjust the ionic strength. One method is to increase the concentration of phosphate buffer from 10 to 500 mM. An alternative is to add NaCl to 200 mM phosphate buffer. The gel formed with low ionic strength (24 mM) consisted of a dense network of 15–20 nm wide and 200–250 nm long nanofibrils observed in low resolution transmission electron microscopy (TEM) images. As the ionic strength increases to 261 mM, as the gel strength and opacity increase, the fibril size increases and forms a bundle. Similarly, Harris et al. At 529 mM and 1300 mM ionic strength, Gobeaux et al. found that the polymeric collagen system is biphasic. Specifically, at a 529 mM ionic strength, a large bundle of fibrils is surrounded by nanofibrils (width about 40 nm, length about 600 nm). On the other hand, at 1300 mM ionic strength, the nanofibrils are surrounded by larger fibrils. The authors report a similar structure when ionic strength is adjusted using a monovalent salt.

The method based on non-destructive in situ SHG imaging provides new structural information about the overall microscopic morphology of the synthesized collagen hydrogel. In our study, the thickness of the collagen hydrogel was a few millimeters. The structural information presented in this study could not be obtained by the transmission electron microscopy (TEM) method used in previous studies. TEM imaging is often destructive to biological samples and requires the transmission of electron beams through ultra-thin samples. Unlike previous studies using SHG imaging, we evaluated parameters related to in situ collagen fiber structure without interfering with the material. For example, unlike the study by Li et al., we do not need to apply a sample on a glass substrate to form a thin portion to pass the transmitted light to form an optical image.

To quantify the relative differences in hydrogel microstructure parameters, we used a method to directly measure fiber and effective pore size. Although we can obtain measurement numbers repeatedly, data analysis can benefit from implementing automated methods. Recent automated methods for analyzing 2D SHG images include the work of Bayan et al. and Matteini et al., who used Fourier transform and entropy analysis to obtain information about the orientation of collagen fibers in gels and corneas, respectively. While these methods work well for fine and/or linear fibers, these methods tend to be more comprehensive types of comparisons. It is not clear how to consider the fiber width of the larger/diffuse fibers we observed at lower temperatures. Texture analysis was also used to analyze SHG images of collagen structures in the outer stromal matrix of normal ovarian stroma and high-grade ovarian serous carcinoma cells. Endogenous magnetic fields and currents occur naturally due to the presence of ions in the wound to direct cell migration. Therefore, it is conceivable that developed data and methods can be used to examine the wound healing process. The synthetic collagen aggregation process can be carried out under ion regulation. Manufacturing medical devices for bone tissue repair requires that the ionic regions of the inorganic molecules of hydroxyapatite interact with collagen on the fibril/fiber surface and may also benefit from this research. The proposed research is also important in science and is also relevant in the field of tissue engineering, especially in the case of using ions as therapeutic agents.

In situ second harmonic generation (SHG) imaging allowed us to determine that increasing the phosphate concentration would increase the length of the polymerized collagen fibers at 37 °C, pH 7.4. When an increased concentration of sodium sulfate was added to the 30 mM phosphate buffer used to polymerize the collagen into a hydrogel, the fiber length also increased. On the other hand, the addition of sodium chloride at physiological and higher concentrations did not significantly extend the fibers in the hydrogel at 37 °C, pH 7.4. Decreasing the temperature produces a multilateral response, depending on the type of ion used. At all temperatures, very high concentrations of fiber length and width of all salts are reduced. The traditional turbidity measurement provides a time scale for the gelation process examined, and the solubility data shows 85% collagen precipitation under all experimental conditions. Ions are important factors to consider when gelatinizing collagen and other biologically derived proteins.

High resolution reflection imaging deep tissue. To demonstrate that our system is capable of producing high-resolution structural images deep in highly scattered tissue, two-photon images are obtained at different depths z, entering the RAFT range from 0 to 230 μm, which is the working distance of the microscope objective. Figure 5.3a–f show two-photon images of RAFT collagen at depths of 0, 50, 100, 150, 200 and 230 μm for λ_{ex} = 800 nm, which is the optimal excitation wavelength for SHG imaging. Obviously, detailed structural information can be derived from the image over the depth of the inspection. The signal gradually decreases as the

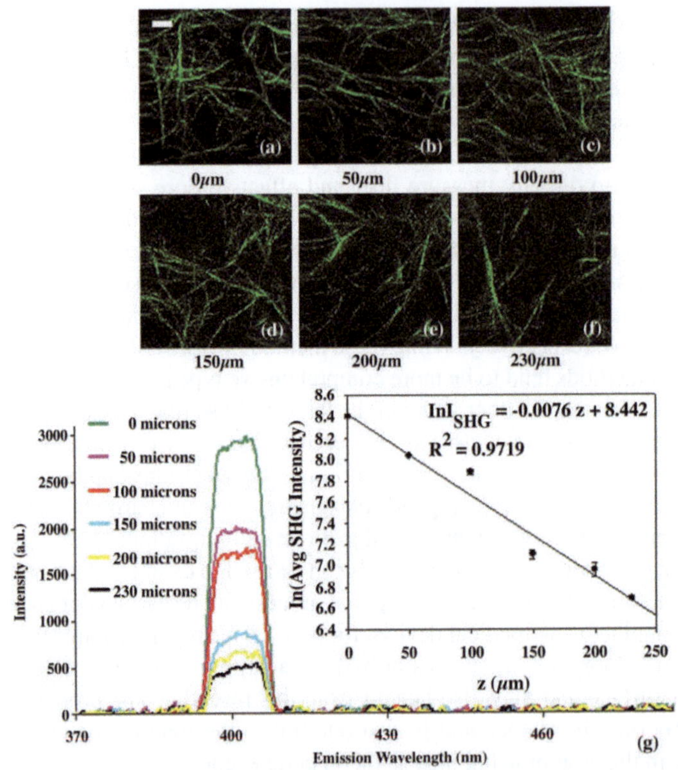

Fig. 5.3 For $\lambda_{ex} = 800$ nm and P = 60 mW (**a–f**), SHG images from RAFT collagen were at depths of 0, 50, 100, 150, 200 and 230 μm. (Bar = 5 μm.) The corresponding spectrum of t = 60 s is expressed in g. (g illustration) Logarithmic plot of SHG intensity versus depth, z (μm)

depth increases, and less collagen fibers can be resolved in the image. However, the maximum imaging depth achieved here is limited only by the working distance of the microscope objective and the low value of the excitation power used (about 5 mW at the sample position). Previous studies using SHG imaging in reflection geometry to obtain structural information into the tissue were characterized by long acquisition times (0.5–2.8 h), even with a sample with an excitation power of up to 80 mW and poor resolution.

To investigate the dependence of SHG intensity on imaging depth in turbid tissue, the spectra corresponding to the images shown in Fig. 5.4a–f were obtained and are shown in Fig. 5.3g. The inset in Fig. 5.3g shows a plot of the SHG peak intensity of the spectrally detected X before = 800 nm versus the natural logarithm of the depth z. The SHG intensity at each depth in Fig. 5.3g is the average of five measurements at different x-y positions in the same z-plane. The experiment was repeated three times to ensure reproducibility of the results. The SHG signal decays with depth due to absorption and scattering in the sample. Multiple light scattering reduces the intensity of ballistic excitation photons reaching the focal region, limiting the number of SHG photons produced in the tissue. The SHG intensity I octave is attenuated as a function of depth z, according to: $(I^{z}_{SHG}) = (I^{z=0}_{SHG}) \exp(-Az)$, where A is the attenuation coefficient for sample absorption at excitation and emission (SHG) wavelengths And a function of the scattering properties. The reciprocal of A produces an attenuation length of l_{att} =132 μm, which is the composite optical property of RAFT and approximates l_{scat} = 137 μm, which is estimated by using the spectral dependence of the SHG intensity. These findings indicate the potential use of SHG depth-dependent decay and the spectral dependence of SHG intensity as a sensitive tool for obtaining quantitative tissue structure information. However, further work is needed to determine the exact physical meaning of these composite parameters.

5.2.2 SHG Imaging for Elastic Arteries

The mechanical properties of blood vessels regulate a variety of phenomena, including pressure, flow, pressure and mass transfer, which in turn have a major impact on cardiovascular function in health and disease. Vascular mechanical energy in microstructured wall components such as collagen and elastin fibers, as well as smooth muscle cells. These microstructured components have different mechanical properties and are subjected to loads of different stress levels. Therefore, blood vessels are known to exhibit complex and unpredictable mechanical behavior. Previous attempts to determine morphological features such as diameter, length, number density, orientation, and curvature of collagen and elastin fibers have typically employed destructive differential digestion techniques. These methods fail to provide useful data because the fibers are very dense and difficult to measure. The main reason for the lack of progress in this area is the lack of a technique that produces three-dimensional rendering of collagen, elastin and smooth muscle cell structures.

Multiphoton microscopy (MPM) is a bioimaging technique that relies on nonlinear photon interactions to provide high contrast and optical slicing capabilities. There are two main types of multiphoton microscopy: second harmonic generation (SHG) and two-photon excitation fluorescence (TPEF). Both types of nonlinear interactions occur in biological tissues without the addition of exogenous contrast agents. Two-photon excitation fluorescence has been widely used to image cells and tissues [7]. Second harmonic generation has recently been used for bioimaging applications. The combination of TPEF and SHG has been used in cell research, thin tissue sections, and more practical thick, unstained Living specimens . Collagen is a well documented source of tissue SHG and autofluorescence. Elastin is also an important source of autofluorescence in extracellular matrices. Collagen and elastin are important determinants of vascular mechanical properties. Their selective visualization is fundamental to determining the microstructural cause of mechanical properties. Fluorescence emission can provide a possible method for separating tissue components based on differential spectral features. However, in the case of collagen and elastin, the emission spectra overlap significantly, thus making their characterization a daunting task. Here, we use TPEF and SHG in series to achieve selective visualization of structural components of the elastic and muscular arteries. We also combined coronary artery dilatation with simultaneous determination of collagen, smooth muscle cells and elastin structure. Specifically, we used a combination of TPEF and SHG microscopy to selectively monitor structural changes in collagen, elastin, and smooth muscle cells in response to different loading conditions. This approach will establish a microstructural basis for observed vascular mechanical properties and demonstrate the potential of MPM as a non-invasive technique for vascular physiology and pathology characterization. The rabbit arteries are highly elastic, and their main components are collagen and elastin in adventia and media, respectively. The content of rabbit elastic artery smooth muscle cells is relatively rare. Images and corresponding spectra of the same sample sites from the same focal plane were obtained from the outer membrane (Fig. 5.4a–d) and medium Fig. 5.4e, f, g, and etc.) The excitation wavelength of the rabbit aortic wall is 800 nm. Wideband filters (320–654 nm) are used to capture TPEF and SHG signals, while 400–465 nm and 520–620 nm bandpass filters are used to acquire SHG (green coded image) and TPEF (red) as color coded images, respectively) . In the outer membrane (Fig. 5.4a–d), collagen is the main component, and the image shows a strong SHG signal from collagen (Fig. 5.4b) and a weak TPEF signal from elastin (Fig. 5.4c). The latter cannot be distinguished in images obtained using wide emission filters (Fig. 5.4a). The corresponding spectrum (Fig. 5.4i, red) shows the 400 nm SHG signal from collagen and the wide TPEF characteristics spanned Due to elastin autofluorescence (excitation/emission maxima 410/500 nm), this region is 410–600 nm with a maximum of 495 nm (inset in Fig. 5.4i). MPM images and spectra show that for the excitation wavelength of 800 nm, the signal from collagen is only SHG, while TPEF is derived from elastin, a molecule similar to a random coil configuration, and therefore does not generate SHG signals. We have confirmed the collagen and elastin structures shown in the images using conventional histology (Fig. 5.4j). In the medium (Fig. 5.4e–h), we observed a strong TPF signal from elastin (Fig. 5.4g) and a weaker SHG signal from collagen (Fig. 5.4f), these signals were not recognized in Fig. 5.4. The corresponding spectrum (Fig. 5.4i, blue) shows a 400 nm SHG signal with lower intensity than that obtained. From the outer membrane (Fig. 5.4i, red), the fact that the collagen fibers in the medium are less is reflected. However, the TPEF signal from the medium is significantly higher, corresponding to a larger elastic fiber density in the medium. The morphology of the MPM images obtained from the medium was consistent with the aortic histology shown in Fig. 5.4j. Cover the green image (Fig. 5.4b and f, respectively, the outer and middle) and the red image (Fig. 5.4c and g, respectively, the outer and middle), producing a high-contrast image showing the outer membrane (Fig. 5.4d) And the structural details of the medium (Fig. 5.4h). The main effect of immobilization is that the elastin fibers in the fixed sample appear to be curved compared to the more elongated straight elastin fibers observed in fresh samples. Parassasi et al. have performed similar observations on aortic fixation. The reason why elastin fibers obtain a curved shape in a fixed sample is that elastin cannot be immobilized.

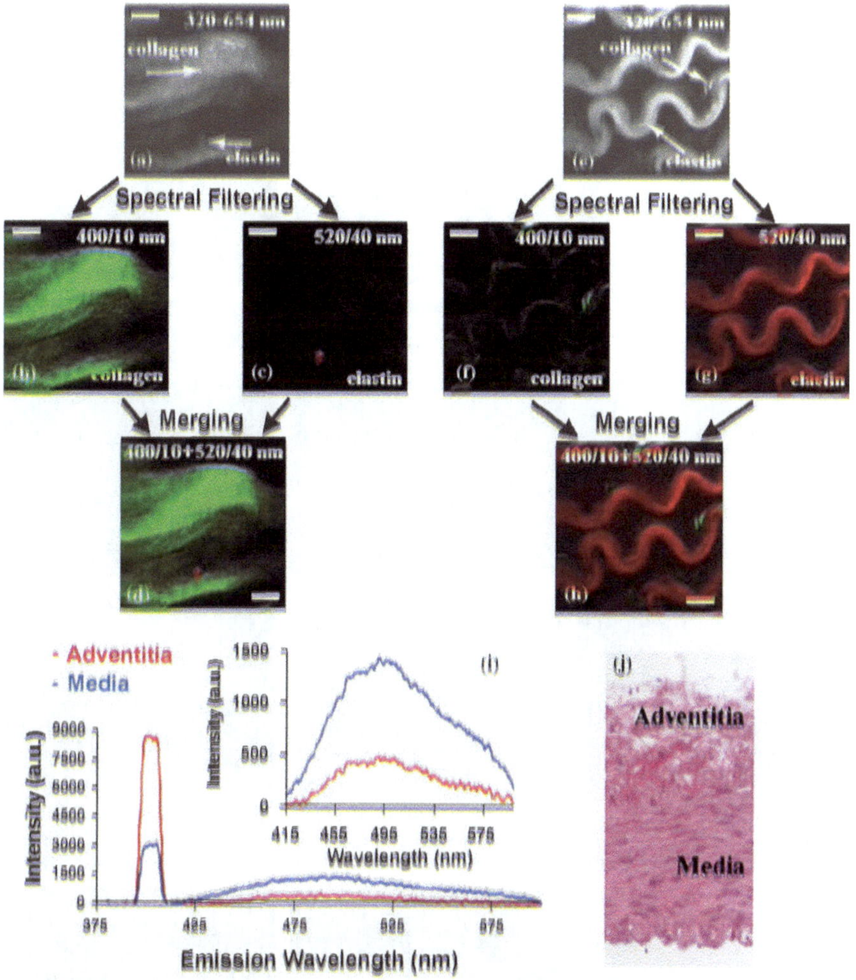

Fig. 5.4 Characterization of MPM signals from elastic arteries. MPM images obtained from the same part of the rabbit aortic wall adventitia using various emission filters: SBG39 emission filter (320–654 nm) (**a**), 400/10-nm filter (**b**) and 520/40-nm bandpass transmit filter (**c**). The superimposed image of b and c is displayed in d. MPM images obtained from the same focal plane in rabbit aortic media using SBG39 emission filter (**e**), 400/10-nm filter (**f**) and 520/40-nm bandpass emission filter (**g**) . The superimposed image of **f** and **g** is displayed in (**h**). The scale is 5 mm. The emission spectra corresponding to **a–d** and **e–h** are shown in (**l**) in red and blue, respectively. The illustration in **l** shows only the TPEF spectrum. Hematoxylin and eosin stained sections of rabbit aortic wall (**J**)

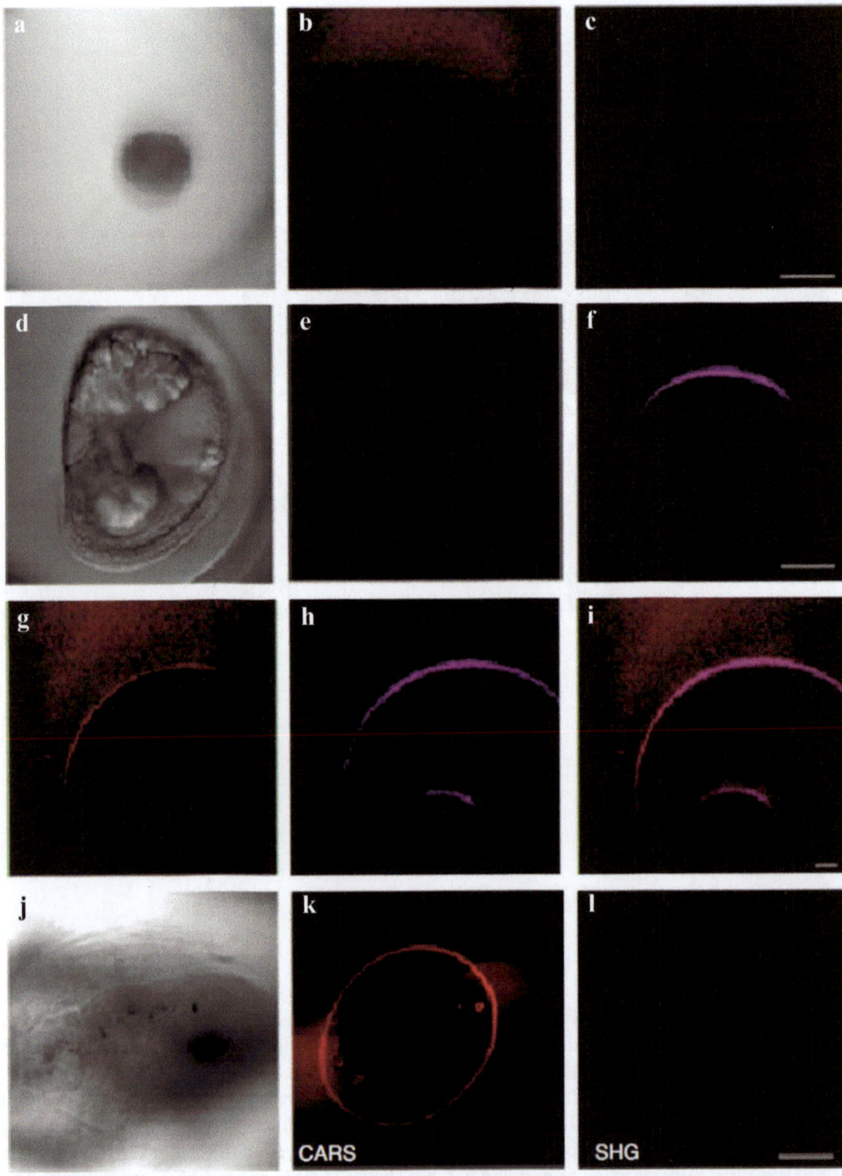

Fig. 5.5 CARS (red) and SHG (pink) images of snail evolution dynamics from day 1st to 9th

5.2.3 SHG Imaging for Snail

Snails refer to all terrestrial species of Gastropod. In Chinese, snails only refer to terrestrial species, and snails in a broad sense also include giant shield slugs. The snail is an animal that includes many different families and genera. Feeds on plants and lays eggs in the soil or on trees. It is more common on tropical islands, but some also live in cold regions. The arboreal species are bright in color, while the terrestrial ones usually have several similar colors, usually with stripes. The crystal snails in Africa are the largest, more than 20 cm in length. Several species of the snails in Europe are often used as delicacies, especially in France. Snails are the most common molluscs on land, and they have high edible and medicinal values. The structure of the snail is mainly a shell made of calcium carbonate and a soft body. Due to the different species, the structure of each species of snail is slightly different. Snails generally live in relatively humid places, where there are plants to avoid direct sunlight, and no one will interfere. Many snail species can be found in areas such as miscellaneous woods and virgin forests; even in the corners of flower beds or gardens, there are opportunities to find snails. Snails that live in cold regions will hibernate, and species that live in the tropics will also dormant in the dry season. The mucus secreted during dormancy forms a calcareous film to seal the mouth of the shell. The whole body is hidden in the shell and connected by collagen. Therefore, nonlinear optical microscopy can characterize the formation of calcium carbonate bone and the growth of collagen during the growth of snails. Li et al. [82] conducted a dynamic imaging study on it, see Fig. 5.5. It can be seen from the figure that calcium ions are continuously enriched by collagen (SHG) from the embryo to the forming process of the snail, and eventually become an adult snail.

Chapter 6
The Principle, Application and Imaging of TPEF

6.1 Principles of TPEF

6.1.1 Two-Photon Absorption

Simultaneous two-photon absorption (TPA) is a nonlinear optical process. This phenomenon was first predicted by Gppert-Mayer in 1931 who calculated the transition probability for a two-quantum absorption process. Two-photon absorption can create excited states with photons of half the nominal excitation energy, which can provide improved penetration in absorbing or scattering media. TPA is a third-order nonlinear optical process. Two-photon absorption is the I^2 dependence of the process, which allows for excitation of chromophores with a high degree of spatial selectivity in three dimensions through the use of a tightly focused laser beam. Theoretical analysis of TPA have been implemented to interpret experimental measurements of TPA, which is very useful to understand structure-property relations. Coupled with finite field techniques, the ab-initio methods are widely used to calculate off-resonant NLO responses, more general method is the time dependent perturbation theory. The sum-over-states (SOS) model is an essentially practical methods, involving calculating both ground and excited-states wavefunctions and the transition dipole moments between them. The transition dipole moment in TPA includes two parts as three-state and two-state models, respectively. Richter, M., and S. Mukamel et al. demonstrated that the two-photon absorption process is a two-step process in which an intermediate state must be experienced during two-photon excitation. The complete two-photon absorption probability is the sum of the processes that traverse all intermediate states:

$$\delta_{tp} = 8\sum_{\substack{j\neq g \\ j\neq f}} \frac{|\langle f|\mu|j\rangle \langle j|\mu|g\rangle|^2}{\left(\omega_j - \frac{\omega_f}{2}\right)^2 + \Gamma_f^2}\left(1 + 2\cos^2\theta_j\right) + 8\frac{\left|\Delta\mu_{fg}\right|^2 |\langle f|\mu|g\rangle|^2}{\left(\frac{\omega_f}{2}\right)^2 + \Gamma_f^2}\left(1 + 2\cos^2\phi\right) \quad (6.1)$$

M. Sun et al., *Linear and Nonlinear Optical Spectroscopy and Microscopy*,
Progress in Optical Science and Photonics 29,
https://doi.org/10.1007/978-981-99-3637-3_6

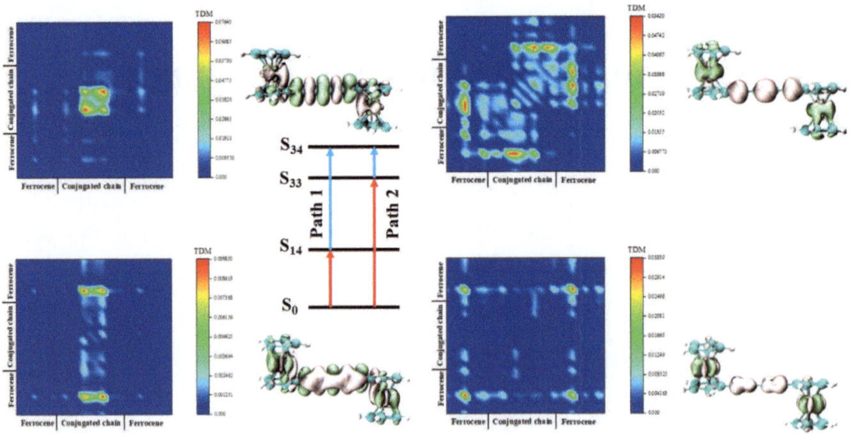

Fig. 6.1 Schematic diagram of charge transfer in a two-step two-photon absorption process

Therefore, if you want to analyze the specific transition behavior in the two-photon absorption process, then the analysis by the two-step transition model is the most suitable for the physical scenario. Figure 6.1 shows the two excitation behaviors when the intermediate states are different in a two-step transition. The left side of the figure is the first charge transfer excitation and the local excitation, and the right side of the figure is the two-step charge transfer excitation (Fig. 6.2).

Based on the analysis and discussion of three molecules, some laws about charge transfer in long-chain molecules were discovered. Firstly, the molecular conjugated chain does enhance the degree of charge transfer. For the molecules discussed in this paper, as the molecular conjugated chain becomes longer, the degree of charge transfer becomes stronger. Secondly, as the molecular conjugated chain grows longer, the contribution of ferrocene on both sides of the molecule to charge transfer becomes weaker. Finally, the multi-center orbital in the conjugated chain can significantly promote the red shift of the charge transfer absorption peak. In order to analyze the two-photon excitation process, this paper proposes an analysis method that best fits the physical picture in the two-photon transition process, that is, the method that can analyze the excitation behavior of the intermediate state. And this method is universal for a variety of molecular structures, including symmetric and asymmetric molecules.

Response Theory

The two-photon cross section σ is given by the factor $(2\pi e)^4 \nu_\mu \nu_\lambda$ times a normalized line shape function $g(\nu_\mu + \nu_\lambda)$ times the square modulus of a certain sum S_{of} over all molecular states:

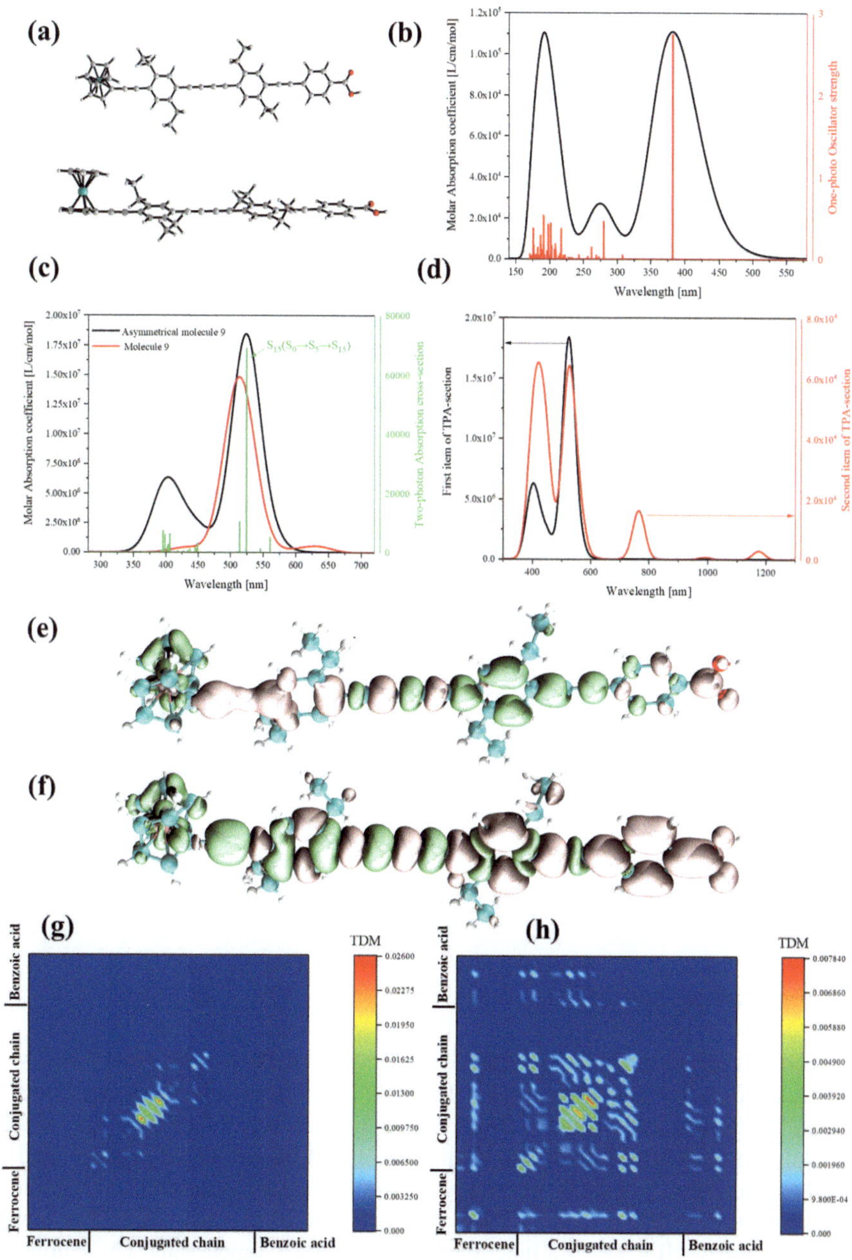

Fig. 6.2 Two-photon absorption transition behavior in an asymmetric system

$$S_{of}(\lambda,\mu)=\sum_{l}\left(\frac{(\lambda\cdot\mathbf{P}_{ol})(\mathbf{P}_{lf}\cdot\mu)}{k_l-k_\lambda}+\frac{(\mu\cdot\mathbf{P}_{oi})(\mathbf{P}_{if}\cdot\lambda)}{k_i-k_\mu}\right) \tag{6.2}$$

where $\boldsymbol{\lambda}=(\lambda_X,\lambda_Y,\lambda_Z)$ is the polarization vector of photon λ in lab coordinates X, Y, Z, $\mu=(\mu_X,\mu_Y,\mu_Z)$ is the same for photon μ_l.$\mathbf{P}_{oi}=\left(L^l_\xi, L^l_\eta, L^l_\xi\right)$ is the transition moment between states o and i written in molecular coordinates ξ,η,ζ, $\mathbf{P}_{lf}=\left(M^l_\xi, M^l_\upsilon, M^l_\zeta\right)$ is the same for states i and f, k_l is the intermediate states energy, and k_λ, k_μ are the photom energies.

The Dalton program requires two input files for calculation. The first file ending in dal controls the calculation method and the keywords of the requirement. The second file ending in mol records the coordinates and basis function information of the molecule.Below is an example of a dal file that uses the response theory to calculate a two-photon absorption spectrum.

```
**DALTON INPUT
.RUN RESPONSE
*PCM
.SOLVNT
H2O
.NEQRSP
*PCMCAV
**WAVE FUNCTIONS
.DFT
B3LYP
**RESPONSE
*QUADRATIC
.TWO-PHOTON
.ROOTS
6
**END OF DALTON INPUT
```

The first keyword ".RUN RESPONSE" means the use of response theory. "PCM", ".SOLVNT" is the key word for a polarizable continuous solvent model. Among them, "H2O" is a water solvent. If you need a different type of solvent, you need to consult the solvent in the manual. If the solvent to be calculated is not included in the procedure, it may be necessary to use a dielectric constant to customize the solvent."**WAVE FUNCTIONS" below to control specific calculation methods and details. "B3LYP" is a functional form in density functionals. The following "**RESPONSE" is the specific form of response theory. You can write ".TWO PHOTON" to calculate two-photon absorption, and ".ROOT" to control how many excited states need to be calculated.

Another mol file records the atomic coordinate information in the molecule. Atoms need to be written in categories, and each class needs to write a base function keyword at the beginning. Of course, you can also use the effective core potential (ECP) here.

```
ATOMBASIS
test molecule
Generated by Multiwfn  // Multiwfn can easily generate Dalton input files
Atomtypes=3 Angstrom Nosymmetry charge=0
Charge=1.0 Atoms=18 Basis=6-31G*
H1     4.51482935     1.52025805    -2.32835266
H2     4.79062647     1.88255336     2.01338342
H3     6.99629437     0.75650079    -1.58418611
H4     3.14788493     2.19269321    -0.10256162
Charge=6.0 Atoms=24 Basis=6-31G*
C1     6.21953128     1.12326562    -0.92649198
C2     4.18803950     1.89759351    -0.14421610
C3     6.30962138     1.24154591     0.49334059
C4     4.90792352     1.52917786    -1.32058582
C5     5.05379955     1.72064117     0.97672477
C6     3.93560052    -1.40003893     1.30368379
Charge=26.0 Atoms=2 Basis=stuttgart_rsc_1997_ecp  ECP=stuttgart_rsc_1997_ecp
Fe1     4.79397278    -0.07179916    -0.01909901
Fe2    -4.79388906     0.07163988    -0.01865695
```

When the calculation is complete, the resulting output file (the file extension name is "out"). The result of the two-photon calculation at the position behind the output file is as follows.

```
***********************************************************************
************ FINAL RESULTS FROM TWO-PHOTON CALCULATION ************
***********************************************************************

The two-photon absorption strength for an average molecular
orientation is computed according to formulas given by
P.R. Monson and W.M. McClain in J. Chem. Phys. 53:29, 1970
and W.M. McClain in J. Chem. Phys. 55:2789, 1971.
The absorption depends on the light polarization.
A monochromatic light source is assumed.

All results are presented in atomic units, except the
excitation energy which is given in eV and two-photon cross
section which is given in GM. A FWHM of 0.1 eV is assumed.

Conversion factors:
1 a.u. = 1.896788 10^(-50) cm^4 s/photon
1 GM = 10^(-50) cm^4 s/photon
+-------------------------------+
| Two-photon transition tensor S |
+-------------------------------+
--------------------------------------------------------------------------------
Sym  No  Energy         Sxx         Syy         Szz         Sxy         Sxz         Syz
--------------------------------------------------------------------------------
1   1    9.97    -0.1     3.2    -3.0    -2.2     0.8     1.2
1   2    9.97     3.5    -2.5    -1.0     0.5     1.5     2.1
1   3    9.98     3.2    -0.6    -2.5    -0.7    -1.6    -2.3
1   4   11.16    -4.0    14.8   -10.6    -9.0     3.7     5.1
1   5   11.16   -13.1     8.8     4.7    -1.3    -7.0   -10.3
1   6   11.17    14.2    -1.5   -12.5    -3.9    -5.7    -8.1
1   7   11.26    -0.2    -7.0     7.0    -9.8     1.7    -1.3
1   8   11.26     0.2     1.1    -1.3     1.9    10.0    -6.8
              ......
1  57   20.88    11.2     1.0   -12.3    -3.6    -6.4   -12.8
1  58   20.88    -0.8    12.3   -11.5   -11.7     7.5     5.1
1  59   20.89    -0.1    13.9   -13.8    14.4    -9.5     8.2
1  60   20.89    -5.8    -1.6     7.3   -11.4   -17.4     7.3
-------------------------------------------------------------------------
```

In this part of the output file, the program first gives the original literature of the response theory. Users should quote these documents at work. Next, give a premise and assumptions. The output of the program is in atomic unit. And will give the conversion relationship of international units. After this information is the information of the two-photon. First, it is the two-photon transition tensor $\hat{S}_{\mu\nu}$. Corresponding in the table are the symmetry number, the excited state number, the excitation energy, and the matrix of the transition tensor representing the matrix element. This tensor is a symmetric tensor. Tensor is an important parameter for calculating the probability of two-photon absorption.

```
                    Transition probabilities (a.u.)
---------------------------------------
D  =  2*Df + 4*Dg, Linear   polarization
D  = -2*Df + 6*Dg, Circular polarization
Df = sum(i,j){ S_ii * S_jj }/30
Dg = sum(i,j){ S_ij * S_ij }/30

Two-photon cross sections
-------------------------------------------------
sigma  =  8*pi^3*alpha^2*hbar/e^4 * E^2*D   (a.u.)

Polarization ratio
------------------------------
R  = (-Df+3*Dg)/(Df+2*Dg)
```

Second, the program gives a formula for calculating the probability of two-photon transitions using transition tensors. This formula is divided into two cases of linearly polarized light and circularly polarized light. After that, a two-photon absorption cross section and a polarizability can be obtained. Next, the program gives the probability of transition, absorption cross section and polarizability for each two-photon excited state in different polarization directions.

```
+-----------------------------------+
|   Two-photon absorption summary   |
+-----------------------------------+
---------------------------------------------------------------------------------------
Sym  No  Energy  Polarization          Df          Dg           D        sigma        R
---------------------------------------------------------------------------------------
1    1    9.97   Linear        0.215E-06  0.110E+01  0.441E+01  0.321E+00     1.50
1    1    9.97   Circular      0.215E-06  0.110E+01  0.661E+01  0.481E+00     1.50

Rotationally averaged two-photon transition strengths

and rate constants

+------------------+---------------+---------------+
|   Polarization   |    DELTA_TP   |   K (0 -> f)  |
+------------------+---------------+---------------+
|  linear (para)   |  4.40697027   |  0.148221E-19 |
```

```
+-----------------+---------------+---------------+
| linear (perp)   |   5.50871209  |  0.185276E-19 |
+-----------------+---------------+---------------+
|    circular     |   6.61045433  |  0.222331E-19 |
+-----------------+---------------+---------------+
1   2    9.97   Linear      0.528E-05  0.110E+01  0.441E+01  0.322E+00    1.50
1   2    9.97   Circular    0.528E-05  0.110E+01  0.662E+01  0.483E+00    1.50
```

6.1.2 Design of Strong Two-Photon Absorption Cross Section

According to research, effective electron delocalization is the basis of two-photon absorption. A typical two-photon absorption conjugated organic molecule includes three parts: π-conjugated bridge (π-bridge), electron donor (donor, D) And electron acceptor (acceptor, A). In order to increase the two-photon absorption cross section, the design of organic molecules has evolved from bipolar (D-π-A), quadrupole (D-π-A-π-D, A-π-D-π-A) to octopole or even Dendritic chromophore, see Fig. 6.3. However, the complex molecular structure will cause structural distortion, which is not necessarily beneficial to increase the two-photon absorption cross section, and at the same time causes the synthesis steps to be very cumbersome. Therefore, whether simple molecules can be used to achieve effective electron delocalization to promote two-photon absorption has become the focus of discussion. Hu's research team made important breakthroughs in the field of two-photon absorption using eutectic methods. They used a donor molecule (4-styrylpyridine) and an acceptor molecule (tetracyanobenzene) to prepare a (styrylpyridine–tetracyanobenzene, STC) co-crystal [83]. The eutectic is obviously different from the color of the two constituent monomers, showing a light yellow color. The crystal structure shows that the D-A and π-π interactions of the donor and acceptor molecules in the STC co-crystal are alternately arranged along the a-axis, forming a three-dimensional D-π-A spatial network. Spectroscopic analysis shows that STC eutectic has charge transfer in both the ground state and the excited state, showing good electronic coupling. Compared with the traditional multipolar chromophore, the eutectic has better electron delocalization based on the charge transfer between the donor molecule and the acceptor molecule, laying a foundation for the realization of two-photon absorption.

6.1.3 Two-Photon Excited/Emitted Fluoresence

Two-photon fluorescence actually has two modes, namely two-photon emission fluorescence (TPEF) and two-photon induced fluorescence (TPIF), see Fig. 6.4 [84]. The transition process of these two modes is very different. Two-photon emission fluorescence is a two-photon fluorescence that is excited by a single photon and emits two photons after relaxation. TPIF is also called two-photon excitation

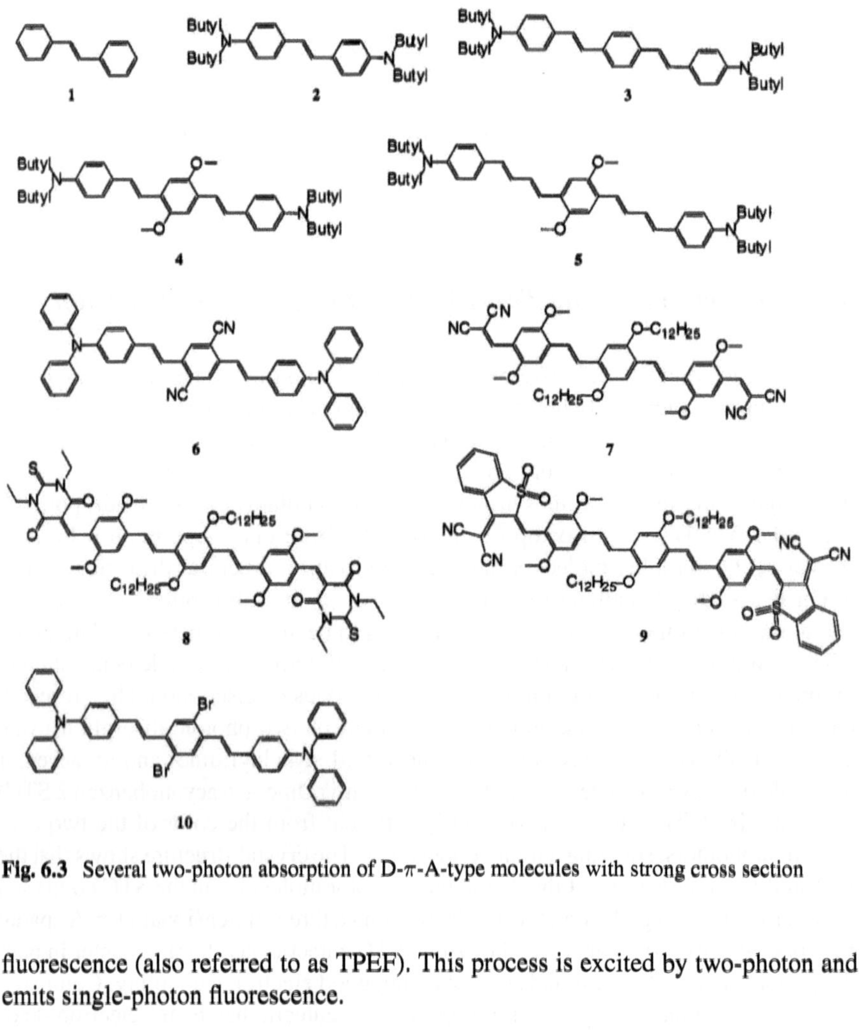

Fig. 6.3 Several two-photon absorption of D-π-A-type molecules with strong cross section

fluorescence (also referred to as TPEF). This process is excited by two-photon and emits single-photon fluorescence.

6.1.4 TPEF Optical Configuration

Herz uses a titanium sapphire laser (wavelength tuning range of 700–1000 nm, pulse width of 100 fs, repetition rate of 80 MHz) and an optical parametric oscillator to form a dual-light source system [85]. When the laser's emission wavelength is 850nm or 920 nm, the optical parametric oscillator The generated wavelength is 1110 or 1170 nm, which can achieve effective excitation of multicolor two-photon fluorescence imaging. Entenberg et al. [86] used two titanium sapphire lasers, one as the excitation light of the fluorophore with a fluorescence absorption wavelength in the range of 750–950 nm; the other laser and optical parameters The oscillator

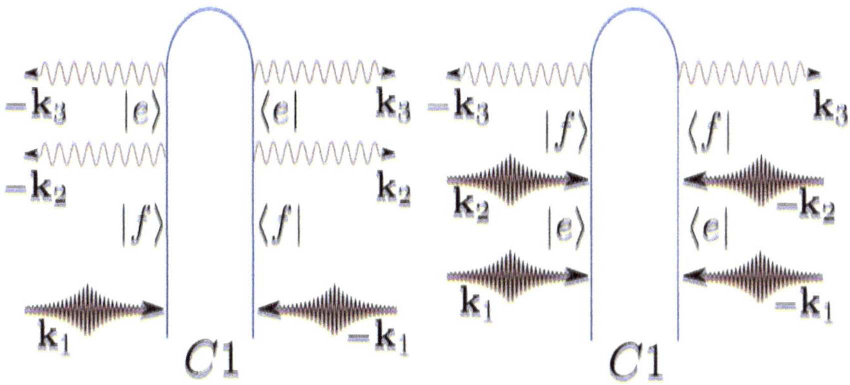

Fig. 6.4 The ladder scheme of two-photon excited/emitted fluoresence [84]

constitutes a new excitation source, which produces excitation light of fluorophores in the range of 950–1040 and 1100–1600 nm.

The raw laser light source realizes multi-color imaging of mouse breast cancer tumor cells labeled with cyan fluorescent protein (CFP), enhanced green fluorescent protein (EGFP) and tagged red fluorescent protein 557 (TagRFP656). Li et al. [87] obtain the technology of optical parametric oscillation frequency doubling ultra-short femtosecond pulses in the 600 nm band, and realizes two-photon excitation in the visible light band. Combined with the traditional 100fs laser, it is used to study the autofluorescence of living bodies, through the detection of NADH, FAD and tryptophan. The two-photon fluorescence imaging has realized the detection of multicolor skin cancer inflammation and the mechanism study of the degranulation effect of mast cells, which is of great significance for early disease diagnosis and pathological research. Mahou et al. used coherent control to achieve The multi-color two-photon fluorescence microscopy imaging is shown in Fig. 6.5. The pulse sequence generated by the titanium sapphire laser and optical parametric oscillation is introduced into the microscope for two-photon fluorescence microscopy imaging, and the second harmonic and third harmonic signals are detected forward. Backward detection of two-photon signal. In Fig. 6.5, τ is the time delay, SHG means second harmonic generation, SFG means sum frequency generation, THG means third harmonic generation, 2PEF means two-photon excitation fluorescence, OPO means optical parametric oscillation, Ti:S stands for titanium sapphire laser, YPF stands for yellow fluorescent protein, td Tomato stands for tomato red fluorescent protein, 1P stands for single photon, 2P stands for two photon. The pulses generated by the titanium sapphire laser and its pumped optical parametric oscillation can maintain coherence Therefore, two pulses of different wavelengths can not only excite the corresponding fluorophores separately, but also achieve mixed two-photon excitation when the time domain overlaps. Using this property, mixed two-photon excitation achieves three-color excitation of different fluorophores. A multi-wavelength light source is the sub-20 fs ultrashort pulse laser.sub-20 fs ultrashort pulse laser can generate

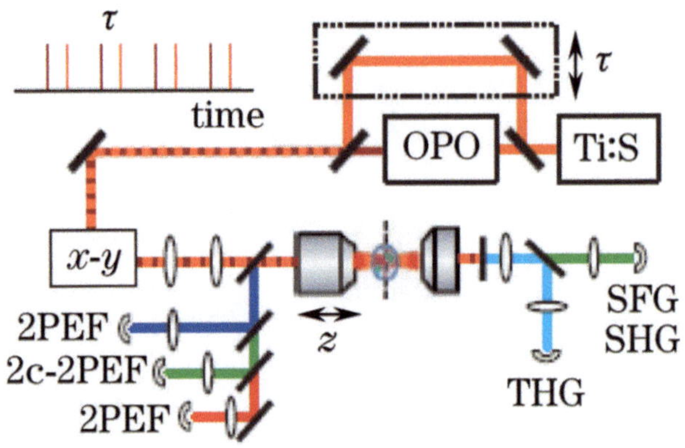

Fig. 6.5 Two-photon fluorescence microscopic imaging device

broadband spectrum that the traditional 100 fs laser does not have, and its spectral width can reach 100–200 nm, covering the excitation of a variety of fluorescent proteins or fluorescent dyes The wideband spectrum of the sub-20 fs ultrashort pulse laser can avoid the complexity and instability of the optical path caused by the collinearity of multiple lasers. Brenner et al. [88] and Pillai et al. [89] use spatial light modulators to control the ultrashort pulse laser pulse The frequency domain phase, combined with the linear de-spectrum method, realizes the selective excitation of multicolor two-photon fluorescence microscopy imaging, and realizes the use of dual detection channels to distinguish three different fluorescent protein-labeled cells. Although multiple fluorophores can be excited at the same time, but The center wavelength of this light source cannot be tuned, and its bell-shaped spectral shape is not conducive to the excitation of fluorescent substances whose wavelength is at the edge of the spectrum, which limits its application in multicolor imaging.

A typical two-photon fluorescence lifetime imaging system is shown in Fig. 6.6 [90]. When the time domain and frequency domain technologies are used for fluorescence lifetime detection, the system uses femtosecond laser and femtosecond laser multiple harmonics as the sample excitation light source [91]. In the system, the excitation light scans the sample plane through a pair of scanning galvanometers to achieve imaging. Specifically, the excitation light passing through the scanning galvanometer is focused on the sample by the objective lens after passing through the dichroic mirror. The excited two-photon excitation fluorescence signal is collected by the same objective lens; then, the fluorescence signal is separated from the excitation light after being reflected by the dichroic mirror, and then enters the detector; finally, the detector inputs the detected signal to the fluorescence lifetime detection module for processing deal with. In terms of frequency domain detection, the fluorescence lifetime detection module measures the phase and modulation of the fluorescence signal relative to the excitation light signal; while in the time domain detection, the

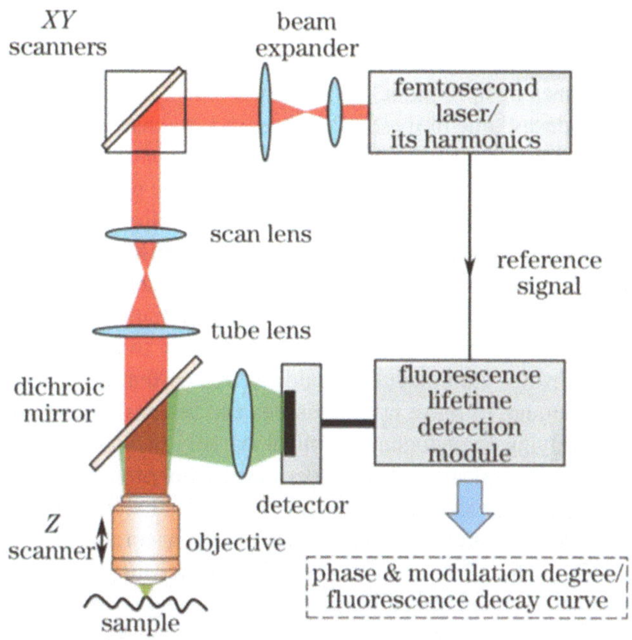

Fig. 6.6 Schematic diagram of a typical two-photon fluorescence lifetime imaging system

module records the decay curve of the fluorescence signal. After processing and analyzing the above measurement results, the system will output specific fluorescence lifetime information and fluorescence lifetime images of the sample.

6.2 Biomedical Imaging of TPEF

A key advantage of two-photon fluorescence lifetime imaging is the ability to reflect the metabolic state of cells and tissues based on autofluorescence, which provides a basis for the technology to be used in tumor diagnosis [92]. Specifically, cell energy mainly comes from sugar metabolism. Normal cells are metabolized by oxidative phosphorylation under aerobic conditions, and metabolized by glycolysis under hypoxic conditions. Unlike normal cells, tumor cells often choose glycolytic pathways as the main way of energy production even under conditions of sufficient oxygen supply. This phenomenon is called the "Warburg effect". NADH and oxidized adenine flavin dinucleotide (FAD), as the main sources of intracellular autofluorescence, play important roles as the main electron donor and acceptor respectively in the process of cell metabolism [3, 93]. It is well known that the ratio of free and protein-bound NADH and the ratio of NADH to FAD can reflect the metabolic state of the cell. The different metabolic methods of normal cells and tumor cells lead to differences in the concentration and state of endogenous optical markers such as NADH and FAD contained in the two, and changes in the state of protein binding and association with cell membranes or lipid particles may cause The fluorescence quantum yield and fluorescence decay time of these substances in vivo have changed significantly. In addition, cytochrome oxidative phosphorylation is controlled by heme. Tumor cells preferentially produce production through glycolytic pathways to significantly reduce their demand for heme compared with normal cells. Therefore, tumor cells can selectively accumulate heme precursors (PpIX). PpIX is also an endogenous optical marker that can produce red fluorescence. But under normal circumstances, the amount of PpIX naturally accumulated in tumor tissue is less, and the fluorescence signal is weak. In imaging, it is usually necessary to introduce exogenously the precursor of PpIX—aminolevulinic acid (ALA), so that PpIX can accumulate in tumor tissues enough to make normal tissues and tumor tissues have better contrast. Therefore, relying on endogenous optical markers such as NADH, FAD, PpIX and other endogenous optical markers to perform two-photon fluorescence lifetime imaging of tissues can reveal the metabolic differences between normal cells and tumor cells, which has certain potential in achieving accurate tumor diagnosis at the cellular level, and has clinical application significance major. At present, NADH and FAD two-photon fluorescence lifetime imaging have been used for the detection of precancerous lesions and cancers, and PpIX two-photon fluorescence lifetime imaging research is also actively carried out at the level of cancer cells and tumor animal models. The tumor detection field covered by the two-photon fluorescence lifetime imaging research based on the above-mentioned endogenous optical markers involves a variety of tumors such as digestive tract tumors, brain tumors, and skin cancers. In the detection of gastrointestinal tumors and brain tumors, although the current research is still very limited, it has huge clinical transformation and application prospects. Skin carcinogenesis, especially melanoma, basal cell carcinoma, etc. has become one of the main detection areas of current two-photon fluorescence lifetime imaging applications, which is supported by relatively large clinical trial data.

6.2.1 *TPEF and Lifetime Imaging for Glioma*

Brain tumors are one of the ten most common cancers in my country. Most of them grow infiltrating and have no obvious boundaries, which makes it difficult to remove completely during surgery and easily damage normal brain functions. Detecting brain tumors and realizing precise identification of their boundaries, so as to completely remove the tumor tissue while preserving the brain functional areas and nerve conduction bundles to the greatest extent, which can effectively reduce the recurrence rate and significantly improve the prognosis.

At present, several exploratory studies based on the two-photon fluorescence lifetime imaging technology to detect brain tumors have emerged in the world. Since 2006, Kantelhardt and others have carried out a series of two-photon fluorescence lifetime imaging studies for glioma boundary recognition from cells, animal models, human isolated tissues to in vivo clinical trials (Fig. 6.7). First, based on endogenous fluorescent markers, they studied tumor cells through cultured glioma cells, normal mouse brain tissue slices, mouse orthotopic glioma tissue, and different types of human glioma samples. Structure and photochemical characteristics of normal brain tissues, solid tumors and tumor boundaries. The results show that two-photon fluorescence lifetime imaging of endogenous fluorescent markers can provide high-resolution microstructures at the cell level of tumors, tumor invasion areas and adjacent normal brain tissues, and rely on fluorescence lifetimes to distinguish tumor tissues from normal brain tissues, as shown in the figure As shown in Fig. 6.7a–c. Further in order to achieve more specific brain tumor recognition, Kantelhardt and others introduced 5-ALA-induced PpIX in the study. Compared with normal tissues, tumor cells can selectively accumulate PpIX, thus having significantly enhanced PpIX fluorescence. Through ex vivo imaging studies on mouse glioma tissue in situ, they found that two-photon microscopy imaging can distinguish tumor cells from adjacent normal brain parenchymal subcellular levels, and compared with normal tissues, tumor tissues are double The photon fluorescence lifetime is significantly longer. This shows that only relying on a quantitative parameter of fluorescence lifetime can define and display brain tumor boundaries [37]. In 2016, Kantelhardt et al. [49] started related in vivo clinical trials. They used the two-photon fluorescence lifetime imaging system developed by Jenlab GmbH-MPTflexTM to perform intraoperative imaging on patients with glioblastoma based on endogenous fluorescent substances. This is the first time that two-photon fluorescence lifetime imaging

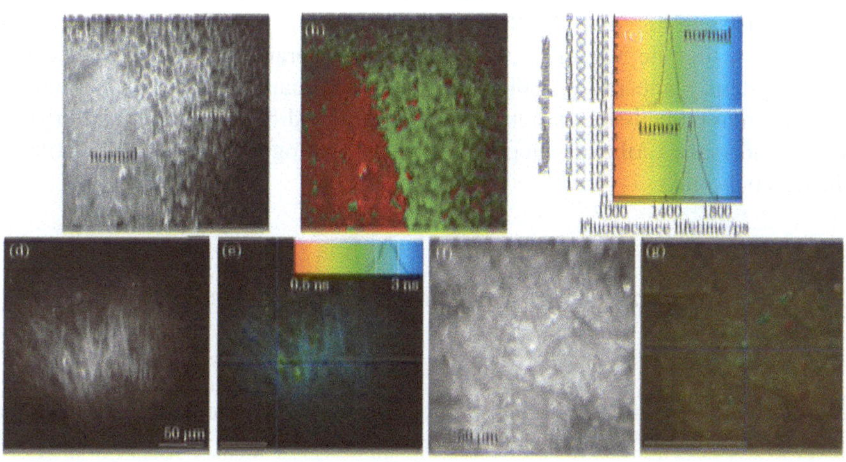

Fig. 6.7 **a–c** Ex vivo imaging of mouse glioma border and **d–g** intraoperative imaging of human glioblastoma. **a** Fluorescence intensity diagram; **b** Fluorescence lifetime coding diagram; **c** Fluorescence lifetime probability distribution curve of normal area and tumor area [34]; **d** Arachnoid fluorescence intensity diagram; **e** Arachnoid fluorescence lifetime coding Figure; **f** Fluorescence intensity diagram of glioblastoma; **g** Fluorescence lifetime coding diagram of glioblastoma

technology has been applied to surgery. Research on imaging of midbrain tumors. They observed arachnoid and solid tumor tissues in vivo, and found that the microstructure and fluorescence lifetime of the two are different, as shown in Fig. 6.7d–g. In addition, Zanello et al. [50] also conducted imaging studies on samples of human meningiomas classified by the World Health Organization as Class I and Class II based on the two-photon autofluorescence signal. Their analysis results show that the detected autofluorescence includes NADH, FAD, lipofuscin, porphyrin, and chlorine signals. Based on the detection of these endogenous fluorescent substances, two-photon microscopy imaging can provide the "gold standard for clinical diagnosis". Histopathological imaging is very similar to the tissue structure. In addition, from level I to level II, the autofluorescence intensity increased significantly, and the fluorescence lifetime also changed significantly. The above research shows that two-photon fluorescence lifetime imaging provides a high-resolution non-invasive optical tissue analysis technology that can output structural and photochemical information at the same time. Relying on this technology, it can distinguish between normal brain tissue and brain tumor tissue. The further development of this technology will likely provide a powerful tool for the identification of intraoperative brain tumor resection boundaries and the detection of residual tumor tissue. However, the metabolic state is a vital feature of the body, and the conclusions obtained by Kantelhardt and Zanello from cultured cells and tissues in vitro cannot be directly derived from the living state; on the one hand, Kantelhardt et al.'s in vivo studies lack the typical normal brain tissue structure and brain. The imaging results of the tumor boundary, on the other hand, the time resolution (200 ps) of the MPTflexTM system is comparable to the lifespan difference (200–400 ps) of the two-photon autofluorescence of brain tumor and normal brain tissue, which cannot meet the needs of brain tumor boundary recognition research. All in all, the current exploration of using two-photon fluorescence lifetime imaging technology to realize brain tumor detection lacks in vivo long-term comparative studies and systematic quantitative analysis of normal and cancerous brain tissues, resulting in unclear tumor differentiation characteristics, which hinders the further development of this technology. Towards the direction of clinical application.

6.2.2 TPEF and Lifetime Imaging for Gastrointestinal Cancer

The digestive tract is an important part of the digestive system, starting from the oral cavity and ending at the anus, including the oral cavity, esophagus, stomach, small intestine (duodenum, jejunum, ileum) and large intestine (cecum, colon, rectum, anal canal) and other parts [38]. Among the top five cancers in my country's morbidity and mortality and the top ten global morbidity and mortality, gastrointestinal tumors account for three [39, 40]. Therefore, the early diagnosis and treatment of gastrointestinal tumors are of great significance.

At present, there have been studies on the application of two-photon fluorescence lifetime imaging technology to the detection of oral cancer, colorectal cancer and other gastrointestinal malignancies, and good results have been achieved. In terms of oral cancer detection, Rück et al. [41] studied the differences between different types of squamous cell carcinoma cell lines and normal oral mucosal cells based on free and protein-bound NADH from the perspective of metabolism. The results showed that, compared with normal cells, the average fluorescence lifetime of cancer cells was longer, the total amount of NADH was reduced, and the ratio of free and protein-bound NADH decreased. In addition to NADH and FAD, two endogenous fluorescent substances that are closely related to cell metabolism, research using other endogenous fluorescent markers is also actively being carried out. For example, Shen et al. [43] conducted a cellular level oral cancer detection study based on endogenous bilirubin, and found that oral cancer cells (>330 ps) have a longer fluorescence lifetime than normal oral glial cells (250 ps). Teh et al. [23] carried out oral cancer detection research at the level of animal models based on keratin, collagen, tryptophan, etc., and extracted tissue structure (fluorescence intensity and three-dimensional spatial distribution of spectrum) and biochemical characteristics information (and fluorescence Life-related parameters) more than ten precancerous lesions and early cancer diagnosis indicators. In the detection of colorectal cancer, Lakner et al. [44] constructed a three-dimensional cavity model of colorectal cancer based on human cloned colon adenocarcinoma cells (Caco-2 cells) in vitro, and demonstrated two-photon fluorescence by imaging NADH signals. Lifetime imaging can well characterize the changes in tissue metabolism. This once again confirmed the potential of this technology for tumor detection. The above studies show that there are differences in the morphological structure of normal tissue, precancerous lesion tissue, and cancerous tissue in terms of two-photon fluorescence intensity and spectral three-dimensional spatial distribution, as well as the intrinsic biochemical characteristics of two-photon fluorescence lifetime related parameters. It is possible to diagnose gastrointestinal tumors such as oral cancer, colon cancer and rectal cancer. However, the application of this technology to the detection of esophageal cancer and small intestine cancer has not yet been reported, and its application to research related to gastric cancer is also very rare.

Li et al. [3] first studied the feasibility of two-photon imaging technology in the diagnosis of gastric cancer based on a two-photon microscopic imaging system based on time and spectrum resolution. The system uses the TCSPC module (SPC-150, Becker & Hickl GmbH) to obtain fluorescence lifetime information, and combines the spectroscopy module and 16-channel high-sensitivity photomultiplier tube (PMT) (PML-16-C, Becker & Hickl GmbH) to achieve fluorescence spectroscopy Probe. The horizontal and vertical resolutions of the system are 280 nm and 1.5 μm, respectively, the time resolution is on the order of picoseconds, and the spectral resolution and spectral detection width are 12.5 and 200 nm, respectively. Through the excitation of 750 nm wavelength, the imaging study of normal and cancerous fresh human gastric mucosa was carried out based on the endogenous NADH signal and the second harmonic signal generated by collagen fibers. The results are shown in Fig. 6.8. In Fig. 6.8: yellow, magenta and big red arrows indicate mucosal epithe-

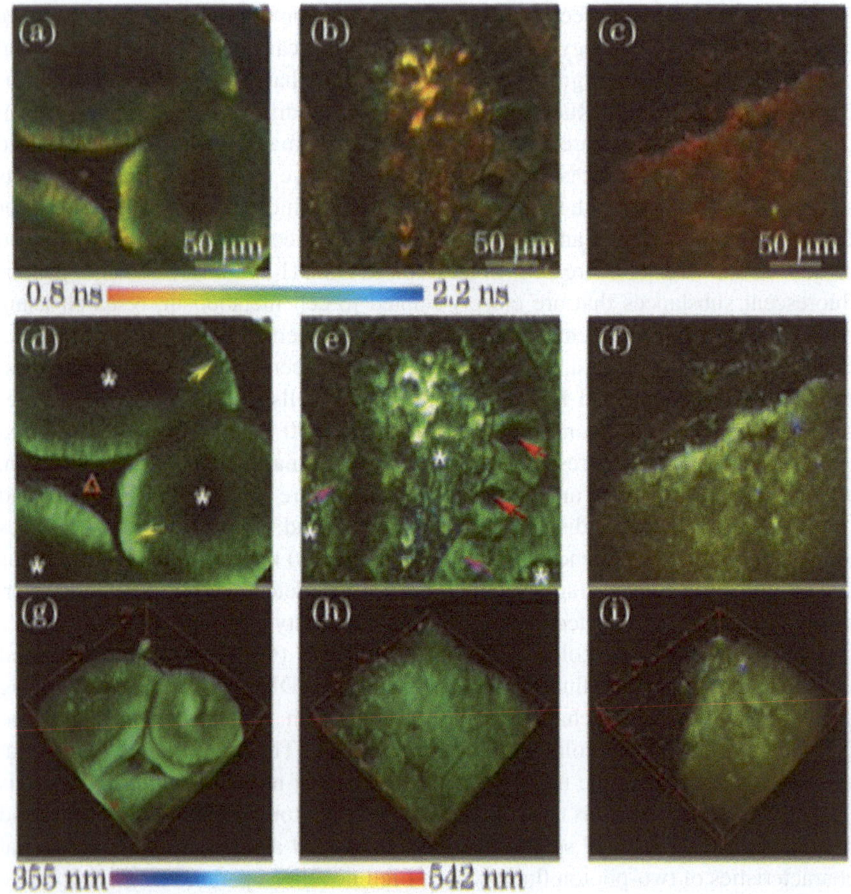

Fig. 6.8 **a–c** fluorescence lifetime coded image and **d–i** fluorescence spectrum coded image of normal and cancerous human gastric mucosa tissue. **a, d, g** normal tissue; **b, e, h** intestinal adenocarcinoma; **c, f, i** neuroendocrine carcinoma

lial cells, cancerous intestinal cells and cancerous goblet cells, respectively; "⋆" and "Δ" indicate interstitial tissue and gastric pits, respectively. Experiments have found that based on the two-photon fluorescence lifetime and spectral information, the mucosal surface structure components such as mucosal epithelium, interstitial tissue, and gastric pits can be clearly distinguished, as shown in Fig. 6.8a, d, g. Based on the identification of the above structural components, normal and cancerous tissues show distinct three-dimensional structural characteristics. In normal tissues, the mucous secretory granules distributed on the top of the cytoplasm of mucosal epithelial cells have no fluorescent signal. According to the vacuole-like structure on the top, a single mucosal epithelial cell is clearly distinguishable. The interstitial tissue between the gastric pits is narrow and contains very small amounts of collagen fibers and plasma cells, even invisible in the superficial layers. The gastric pit has a branch-like opening in the shallow layer and an oval shape at the bottom, as shown in Fig. 6.8a, d, g. The epithelium of intestinal adenocarcinoma tissue presents a structure very similar to intestinal epithelium, with cancerous vacuolated goblet cells and intestinal cells. The cytoplasm of intestinal cells showed a strong fluorescent signal, while the nuclear signal at the bottom was weak. The interstitial tissue is obviously widened, in which collagen fibers and a large number of inflammatory cells can be seen. Collagen fibers appear blue-violet in the fluorescence spectrum, and inflammatory cells appear orange and yellow in the fluorescence lifetime and fluorescence spectrum, respectively. The gastric pit becomes obscured, as shown in Fig. 6.8b, e, h. Neuroendocrine cancer tissues completely lost the regular mucosal surface structure, and only densely arranged and indistinguishable tumor cells gathered into clumps, as shown in Fig. 6.8c, f, i. On the whole, compared with normal tissues, the fluorescence lifetime of cancerous tissues is significantly shorter, and the fluorescence spectrum is red-shifted, especially in neuroendocrine cancer tissues (Fig. 6.8). This is closely related to the changes in cell metabolism caused by cancer. The above results show that only relying on endogenous fluorescent substances, time and spectrum-resolved two-photon microscopy imaging technology can reveal the three-dimensional morphological structure and biochemical characteristics of fresh gastric mucosal tissue at the subcellular level. Therefore, it has great potential in the diagnosis of various types of gastric cancer.

Chapter 7
The Principle, Application and Imaging of STED

In a conventional optical microscopy system, due to the diffraction effect of the optical element, the spot light formed on the sample after the parallel incident illumination light is focused by the microscope objective is not an ideal point, but a diffraction spot having a certain size. Samples within the diffraction spot will fluoresce as a result of illumination by the illumination, such that sample detail information within the range is not resolved, thereby limiting the resolution of the microscopy system. In order to overcome the limitations of the diffraction limit and achieve super-resolution microscopy, how to reduce the effective fluorescent light-emitting area at a single scanning point becomes a key. In STED microscopy, the reduction in effective fluorescent luminescence area is achieved by stimulated emission effects. In a typical STED microscopy system, two illuminations are required, one for excitation and the other for loss of light. When the excitation light is irradiated so that the fluorescent molecules in the diffraction spot range are excited, and the electrons transition to the excited state, the loss light causes some electrons outside the excitation spot to return to the ground state in a stimulated emission manner, and the rest is located in the excitation spot. The excited electrons in the center are not affected by the loss of light and continue to return to the ground state by autofluorescence. Since the wavelengths and propagation directions of the fluorescence and autofluorescence emitted during the stimulated emission are different, the photons actually received by the detector are generated by the autofluorescence of the fluorescent sample located in the central portion of the excitation spot. Thereby, the light-emitting area of the effective fluorescence is reduced, thereby increasing the resolution of the system. Another key to achieving super-resolution in STED microscopy is the non-linear effect of stimulated emission and autofluorescence. When the loss light is irradiated at the edge position of the excitation spot such that the electrons in the sample are subjected to stimulated emission, part of the electrons inevitably return to the ground state in an autofluorescence manner. However, when the intensity of the lossed light exceeds a certain threshold, the stimulated emission process will be saturated. At this time, the electrons returning to the ground state by the stimulated emission mode will occupy the majority, and the electrons returning to the ground state by the

M. Sun et al., *Linear and Nonlinear Optical Spectroscopy and Microscopy*,
Progress in Optical Science and Photonics 29,
https://doi.org/10.1007/978-981-99-3637-3_7

autofluorescence mode may can be ignored. Thus, by increasing the intensity of the lost light, more range of autofluorescence within the excitation spot is suppressed, and the resolution of STED microscopy can be improved.

7.1 Principles of STED

7.1.1 Selection of Excitation and Loss Laser Types

Commonly used lasers are pulsed light and continuous light. At the time of STED microscopy, all STED systems were built based on pulsed light sources. The main reasons are as follows: (1) The use of pulsed light sources makes the excitation and loss light have time domain Separability makes the extinction process of stimulated emission loss easier to handle; (2) Since the resolution of STED microscopy increases with the loss of light intensity used, at the same average power, the pulsed light has a ratio Higher peak light intensity for continuous light. However, the use of pulsed light sources also has drawbacks. In order to achieve better extinction effect, the pulse width of the excitation light and the loss light need to be optimized according to the different fluorescent samples used. The typical value is the excitation pulse width $\Gamma < 80ps$ and the loss optical pulse width is about 250 ps. There should also be a certain time delay between the corresponding excitation light pulse and the loss light pulse. Therefore, pulsed light STED systems often require the placement of optics that broaden and synchronize the pulses, making the STED system very complex and expensive. To make the STED microscopy system relatively simple and inexpensive, researchers began to investigate the possibility of continuous lasers for STED microscopy. SWHell et al. found that when the extinction rate of the loss light is much larger than the excitation rate of the excitation light, the separability of the loss light and the excitation light in the time domain will become less important, thus proposing the first continuous The light source acts as a loss-light STED system. They theoretically and experimentally proved the feasibility of the continuous light STED system and achieved the super-diffraction limit resolution of 29–60 nm. Although the continuous light STED system requires a larger average power, its peak intensity is much smaller than that of pulsed light, thus reducing the possibility of unwanted multiphoton excitation of the phosphor. At the same time, the continuous optical STED system can achieve faster scanning speeds than pulsed STED. When continuous light is used as the loss light, the excitation light can be either pulsed or continuous, which greatly expands the selection of the source type of the STED microscopy system. In summary, the types of laser light sources currently available for STED systems are: (1) pulsed excitation light and pulsed loss light; (2) pulsed excitation light and continuous loss of light; (3) continuous excitation light and continuous loss of light.

7.1.2 Selection of Excitation and Loss Wavelengths

In a typical STED microscopy system, the choice of the wavelength of the excitation and loss light needs to meet the following principles: the excitation wavelength should be chosen near the peak wavelength of the excitation spectrum of the phosphor used to ensure better absorption; At the long-wave tail of the emission spectrum of the phosphor used, to avoid the secondary excitation of the lost light to the sample. However, at this wavelength selection, the stimulated emission cross-section σ at the wavelength of the loss light is small, so that the corresponding larger threshold intensity I_s, resulting in a higher loss of light required, the bleaching of the sample is more serious. To solve this problem, researchers began to consider the loss of light wavelength near the peak wavelength of the emission spectrum to increase the stimulated emission cross section. The key issue at this time is how to eliminate the interference of the fluorescence generated by the loss of light on the secondary excitation of the sample on the experimental effect. Both theory and experiment show that this secondary excitation has a great negative impact on the resolution of the system. Recently, two methods for eliminating the interference caused by loss of light to excite fluorescence have been proposed by researchers. In the first method, only the loss source is turned on, and the fluorescence intensity of the loss spot excited on the sample is detected, and then the excitation light source and the loss light source are simultaneously turned on, and the fluorescence intensity of the conventional excitation is obtained to obtain a final image. The second method is mainly based on the principle of frequency detection. The excitation light is modulated at a certain frequency without modulating the loss light, and the fluorescence component of the modulated frequency is extracted by the lock-in amplifier to eliminate the influence of the loss of light. Although both methods have proved feasible in the experiment, the current development and application are still immature. The first method is complicated by the need to switch the opening and closing of the light source. At the same time, the bleaching, flickering and fluctuation of the noise level of the sample between adjacent detections will have an impact on the final experimental results. In the second method, the use of a lock-in amplifier limits the imaging speed to a certain extent, and the bleaching of the sample by this method is still serious. Therefore, in the current mainstream STED microscopy system, the wavelength of the loss light is still selected at the long-wave tail of the emission spectrum. Improvements in the wavelength of lossy light are yet to be studied.

7.2 Biomedical Imaging of STED

7.2.1 Nervous Structure Imaging

In order to distinguish between the pools of synapto-binding proteins that remain on the surface of the plasma membrane and those that are internalized by endocytosis,

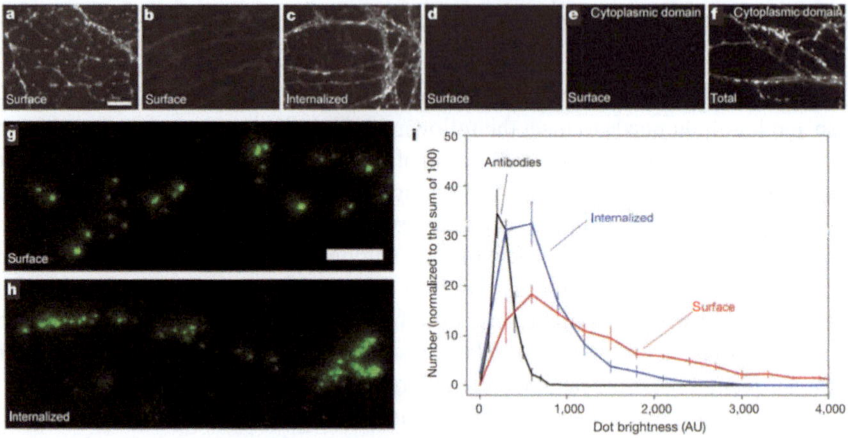

Fig. 7.1 Comparison between surface-exposed and internalized pools of synaptotagmin shows that the protein remains clustered in the presynaptic plasma membrane

we used two different schemes. First, the labeling was performed on ice without Ca^{2+} (i.e., no endocytosis), and bright staining was observed (Fig. 7.1a). Secondly, labeling was performed at 37 ± 8 °C to allow uptake, resulting in antibody binding to the surface exposed and intrinsic synapto-binding protein, and the staining intensity of the non-permeabilized preparation hardly exceeded the background (Figure 7.1b). In contrast, strong labeling was observed after permeabilization, indicating that internalized (unclosed) vesicles can be used for secondary antibody labeling (Fig. 7.1c). After the neurons were labeled with antibodies specific to the cytoplasmic tail of synaptic markers. The labeling can only be observed after the culture has been infiltrated before the antibody incubation (Fig. 7.1e, f). The comparison of surface exposure and internal synapse binding protein pool analyzed by STED microscope is shown in Fig. 7.1g, h. Both internalization and surface exposure pools are resolved as closed points. Therefore, post-synaptic chlorophyll is still concentrated in small clusters after exocytosis, rather than scattered on the plasma membrane. In addition, compared to the internalized pool, fewer spots are always detectable on the surface, which suggests that the plasma membrane pool is short-lived due to rapid endocytosis. It is worth noting that the dots on the cell surface look brighter than the dots on the internalized pool. Therefore, we quantified the brightness of the points (see methods). For comparison, a dilute solution of the primary antibody was adsorbed on the glass, labeled with the secondary antibody, imaged by STED, and quantified in parallel. The resulting histogram (Fig. 7.1i) shows that the internal and surface exposed dots are much brighter than a single antibody, which confirms that each dot represents multiple synaptic small molecules, and the brightness of the surface exposed plaques is comparable to that of internal vesicles. The second observation may be due to the short duration of the endocytosis process (2–5 s), which limits the antibody's accessibility to the epitope. On the other hand, exposing the surface

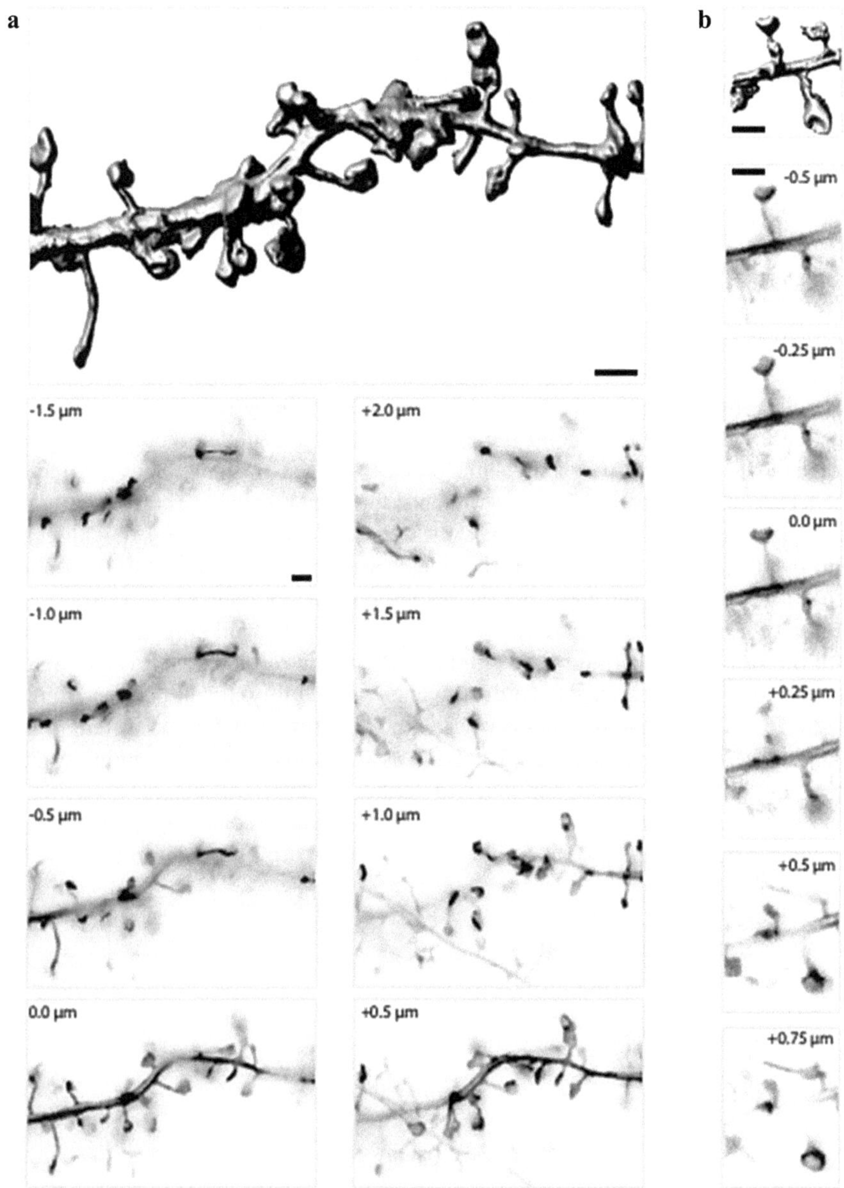

Fig. 7.2 High-resolution 3D STED imaging of dendritic structures. **a** Volume reconstruction of image stack data; the panels show individual image sections acquired at the z level indicated ($\delta z = 0.5\ \mu m$). **b** Another example of a reconstructed stretch of dendrite ($\delta z = 0.25\ \mu m$), the lateral pixel size was 29 nm/pixel in both cases. (Scale bars,1 µm.) [94]

patch to the antibody solution for a few minutes (on ice to inhibit the circulation of activity) to obtain higher labeling efficiency. After infiltration, a comparison of the internalized pool with the total pool (unblocked) shows that a large part of the bright spots (surface) are retained, which indicates that the infiltration will not damage the surface clusters.

7.2.2 3D STED Imaging

Compared with conventional optical micrographs, the sharpness of the dendritic crystals and related dendritic spines images in STED has been significantly enhanced, revealing structural details, and reminiscent of the volume reconstruction of serial sections of electron microscopy. Figure 7.2 shows two examples, these examples are based on the dendritic stretched volume rendering of the 3D image stack obtained by the STED microscope. In these experiments, the axial resolution is the resolution of a standard confocal system. However, the axial resolution can also be significantly improved by STED microscopy. The image was acquired within 0–10 μm above the surface of the cover glass, but the resolution was not reduced. It is expected to be used for STED imaging in optically dense tissues. Considering that the oil lens will produce more aberrations due to the mismatch in refractive index, the use of a water lens will help improve the depth penetration ability. It can be seen from the figure that STED imaging can clearly see the dendritic structure on different focal planes. The data on these focal planes are used to reconstruct the neural dendritic structure. This non-contact, non-labeling method is very beneficial for detecting the nerve structure of organisms and even humans. The 3D reconstruction technology provides a good imaging basis for the operation of neurological deformity [94].

Chapter 8
Plasmon Enhanced Nonlinear Spectroscopy and Imaging

8.1 Principles

8.1.1 Plasmon

Surface Plasmon (Surface Plasmon) is a phenomenon in which electrons on a metal surface collectively oscillate under the action of an external electromagnetic field, see Fig. 8.1. It is divided into Local Surface Plasmon Resonance (LSPR) and Surface Plasmon (Surface Plasmon). Polariton, SPPs) two kinds. LSPR is a conductive electron resonance phenomenon generated at the interface of positive and negative dielectric constant materials under the excitation of incident light. At the resonance wavelength, the near-field field strength is enhanced. This near field is highly concentrated in the nanoparticle and decays rapidly as it moves away from the nanoparticle/dielectric interface into the dielectric substrate. The enhancement of light intensity is an important aspect of local surface plasmon resonance. Localization means that LSPR has a very high spatial frequency (sub-wavelength) and is only limited by the size of nanoparticles. Due to the enhancement of the electric field amplitude, the effects based on the enhancement of the amplitude, such as the magneto-optical effect, are also enhanced by the presence of local surface plasmon resonance. LSPR is the basis for the absorption of materials on the surface of many rough metals (usually gold or silver) or the surface of metal nanoparticles, and it is also the basic principle of many color-based biosensor applications. SPP is an electromagnetic wave in the infrared or visible range that propagates along the metal-dielectric or metal-air interface. The SPP wavelength is shorter than the wavelength of the incident light (photon). Therefore, SPP has stricter spatial constraints and higher local field strength. There is a sub-wavelength constraint in the direction perpendicular to the interface. Surface plasmons will propagate along the interface until the energy disappears, including being absorbed by the metal or scattered to other directions (such as free space). The application of surface plasmons makes sub-wavelength microscopy and lithography beyond the limit of diffraction limit. At the same time, it is also

M. Sun et al., *Linear and Nonlinear Optical Spectroscopy and Microscopy*,
Progress in Optical Science and Photonics 29,
https://doi.org/10.1007/978-981-99-3637-3_8

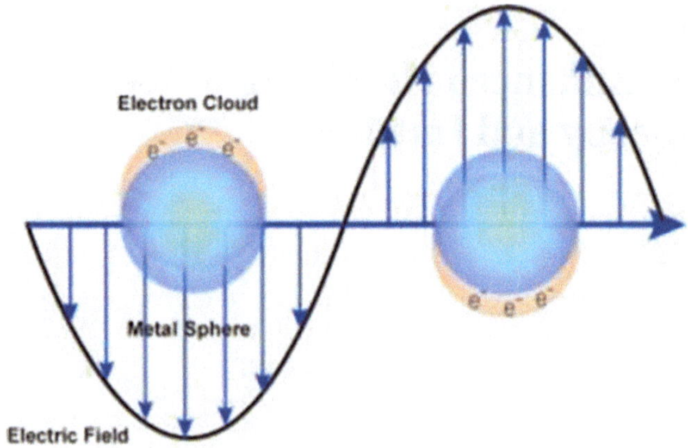

Fig. 8.1 The schematic diagram of physical mechanism for surface plasmon

possible to use the first steady-state micromachine to measure the basic properties of light itself: the photon momentum in the dielectric. Other applications include photon data storage, light oscillation and two-photon effect. Surface plasmons can be excited by electrons or photons. Electron excitation is achieved by emitting electrons into the metal body. As the electrons are scattered, the energy of the electrons is transferred to the plasma. The component parallel to the surface in the scattering vector causes the generation of surface plasmons. For the excitation of surface plasmons with photons, the two have the same frequency and momentum. However, for a given frequency, the momentum of a free-space photon is less than that of SPP, because the two have different dispersion relationships. Momentum mismatch is the reason why photons in free space in the air cannot be directly coupled into SPP. For the same reason, surface plasmons on smooth metal surfaces cannot emit energy into the dielectric in the form of free-space photons (if the dielectric is uniform). This mismatch is similar to the loss of transmitted energy during total internal reflection. However, coupling media can be used to couple photons into surface plasmons, such as using prisms or gratings to match the wave vectors of photons and SPPs (hence momentum matching). In the Kretschmann structure, the prism can be placed close to the thin metal film. In the Otto structure, the prism can be placed very close to the metal surface (as shown in Fig. 8.1). The grating coupler matches the wave vector by increasing the wave vector component in the parallel direction by an amount related to the grating period. This method is crucial for theoretically understanding the influence of surface roughness, but it is rarely used. In addition, simple isolated surface defects such as grooves, slits, or wrinkles on the plane can exchange energy between free-space radiation and surface plasmons, thus creating coupling.

8.1.2 Enhancement Mechanism

The most typical application of localized electric field enhancement of surface plasmons is surface enhanced spectroscopy, which mainly includes surface enhanced Raman scattering (SERS) and surface enhanced fluorescence (SEF). Since the discovery of SERS in the 1970s, SERS has attracted widespread attention and in-depth research; due to its single-molecule detection sensitivity, SERS has very broad application prospects in the chemical and biological fields. From the mechanism, the enhancement of SERS includes the incident light Two parts of enhanced G_0 and enhanced G' of Raman scattered light: [95]

$$G = G_0 G' \approx \left| \frac{E'(\omega_0)}{E_0(\omega_0)} \right|^2 \cdot \left| \frac{E'(\omega_R)}{E_0(\omega_R)} \right|^2 \tag{8.1}$$

where the ω_0 and ω_R are the frequency incident light and Raman scattering light, respectively. If the $\omega_R \approx \omega_0$,

$$G \approx \left| \frac{E'(\omega_0)}{E_0(\omega_0)} \right|^4 \tag{8.2}$$

That is, the SERS enhancement factor is proportional to the fourth power of the electric field enhancement. The electromagnetic enhancement mechanism is considered to be dominant in the SERS enhancement, generally up to $10^5 \sim 10^8$, and theoretical calculations show that under certain circumstances, the contribution of the electric field enhancement can even be Up to 10^{12}, single-molecule SERS detection can be achieved. Generally, if the molecule is directly adsorbed on the metal surface, charge transfer may occur between the molecule and the metal, similar to the process of resonance Raman scattering, which will lead to an increase in the effective polarizability of the molecule, Which also leads to Raman scattering enhancement ($\boldsymbol{P} = \alpha \boldsymbol{E}$), which is generally called the chemical enhancement mechanism, and its enhancement factor is about $10^1 \sim 10^8$ [96]. Due to the effect of electromagnetic coupling, the aggregation of metal nanoparticles will produce additional electromagnetic enhancement. Under suitable excitation conditions, it is much higher than that of individual particles. For example, for the dimer of nanoparticles. When the polarization direction of the incident light is parallel to the long axis of the dimer, the electromagnetic coupling effect is the strongest, while the vertical polarization is the weakest. On the other hand, nanoparticle aggregates can also modulate the polarization and emission direction of Raman scattered light. More complex coupling systems also include the coupling of metal nanoparticles and nanopores, and the coupling of nanoparticles and metal nanowires. As well as the coupling between nanoparticles of different morphologies and different materials. It is worth noting that when the particle spacing in nanoparticle aggregates is about 1–2 nm or even smaller, as the spacing is further reduced, the gap between the particles The electron tunneling effect starts to work, and the electromagnetic enhancement will be partially offset. An intuitive hypothesis is that when the distance between two particles in the dimer is infinitely reduced, the two particles will merge into one and become one with a certain long diameter. Compared with nanorods, the electromagnetic coupling effect will not exist, and the average electromagnetic enhancement effect will be much lower than that of the dimer of nanoparticles. This phenomenon has recently become a hot topic in SERS and surface plasmon research. Because when the particle spacing is less than 1 nm, theoretically, it can no longer be explained by classical electrodynamics, and the concept of quantum must be introduced, that is, quantum plasmonics (quantum plasmonics). In fact, the mechanism of surface plasmons to enhance nonlinear optical signals is very similar to SERS. It also enhances the incident light and outgoing light at the same time. However, due to the existence of the nonlinear optical process, the frequency difference between incident light and scattered light or emitted light is very large, so the approximate conditions are no longer satisfied. Mi et al. [82] recently reported a nanostructure that can simultaneously enhance 800 nm incident light and its frequency multiplication light, which is of great help to the enhancement and measurement of nonlinear optical spectra.

8.2 Application of Surface Plasmon Enhanced Nonlinear Optical Signals

8.2.1 Surface Plasmon Enhanced CARS

Surface plasmons can significantly enhance incoherent and coherent nonlinear processes, including CARS, SHG and TPEF. Relying on plasmonic nanostructures to confine the nano-level light field provides an effective method for enhancing spectral signals and optical imaging. Plasma is caused by the coordinated coherent oscillation of free electrons excited by light at optical frequencies. The synthesis of Au @ Ag nanorods can significantly enhance CARS. The relationship between the physical mechanism and the enhancement factor is shown in Fig. 8.2. For the third-order nonlinear interaction, three incident photons simultaneously interact with the dipole, annihilate and produce a new photon, the frequency of the new photon is related to the frequency of the incident photon. If the frequencies of all incident photons are equal, then the energies of the three incident photons are added together to emit photons at three times the frequency, the process is called third harmonic generation (THG). The first implementation of SHG enhanced surface plasmon resonance (SPR) dates back to the 1970s. On the rough silver surface, surface enhanced SHG appears. The 3D mushroom array model can significantly improve the scattering intensity of plasma-enhanced SHG. Local SPR is considered to be a good method for metal nanoparticles (NP) to enhance the weak SHG of biomolecules, making it widely used in biological imaging.

Immunoscopy is the first application of immunostaining in biological imaging. Prostate biopsy marked white light (left) and immune fragments (right) as p63 antibodies bound to plasma NP, as shown in Fig. 8.3. SECARS signal is a biological

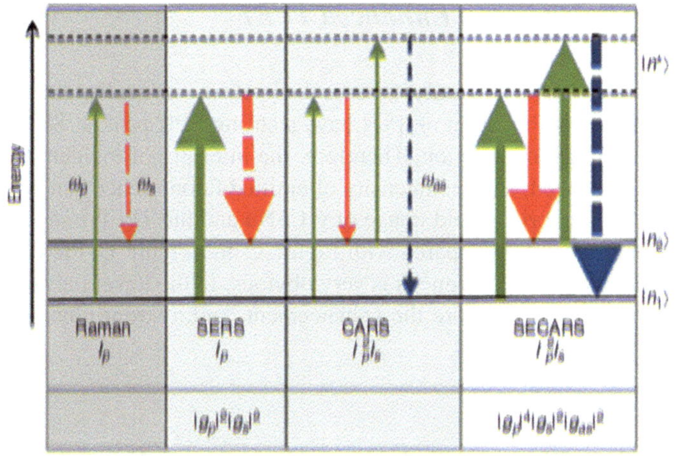

Fig. 8.2 The physical mechanism of surface enhancement CARS

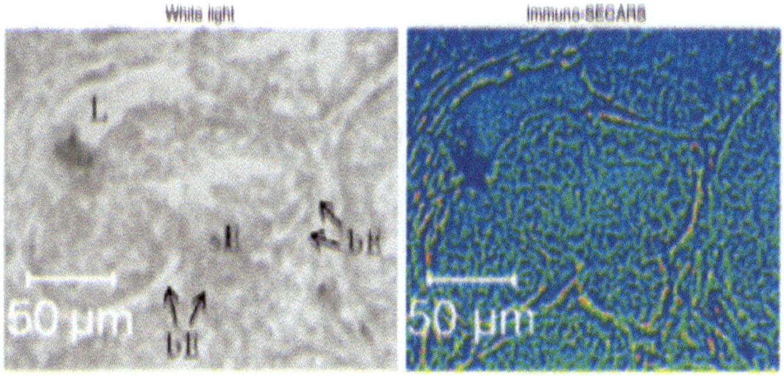

Fig. 8.3 The bright-field and SECARS image of prostate tissue [97]

application, and its signal comes from a label molecule that binds to plasma NPs and interacts with prostate tissue through NPs-binding antibodies. Immunoscopy uses specific target nanoprobe staining, which can be achieved simultaneously with multiple dyes in various spectral modes. This makes it possible to simultaneously image different cellular targets. Various tags can achieve great advantages in specific biological processes. The application of unlabeled SECARS in bioimaging was carried out under a SECARS microscope with a lipid structure. The lipid structure was placed on a 30 nm thick flat gold substrate to present a clear SECARS image of cholesterol oleate.

8.2.2 Surface Plasmon Enhanced TPEF

Surface plasmons can significantly increase the intensity of incident light. The metal nanorods synthesized by Mi et al. [82] have a strong SPR peak at 800 nm and its frequency multiplication position. Therefore, the intensity of incident light during TPEF can be enhanced, and the intensity of emitted fluorescence can be enhanced. Figure 8.4 shows the bright-field image of g-C3N4 and the TPEF image before and after SPR enhancement. Compared with Figure C and Figure D, the enhancement of TPEF signal by surface plasmons is very obvious. In the lower part of the figure, there is almost no signal before the enhancement, and a strong imaging signal is present after the enhancement.

8.2.3 Surface Plasmon Enhanced High-Order Harmonic Wave Generate

When the SPR wavelength of the specific nanostructure matches the incident wavelength of the higher harmonic, the higher harmonic signal can be significantly enhanced. At this time, if there are multiple SPR peaks in the nanostructure, which coincide with the incident and emission in the high-order harmonic process, the high-order harmonic signal can be enhanced synergistically. Figure 8.5 shows the SHG and THG imaging images of nerve cells. It can be found that there is a high signal

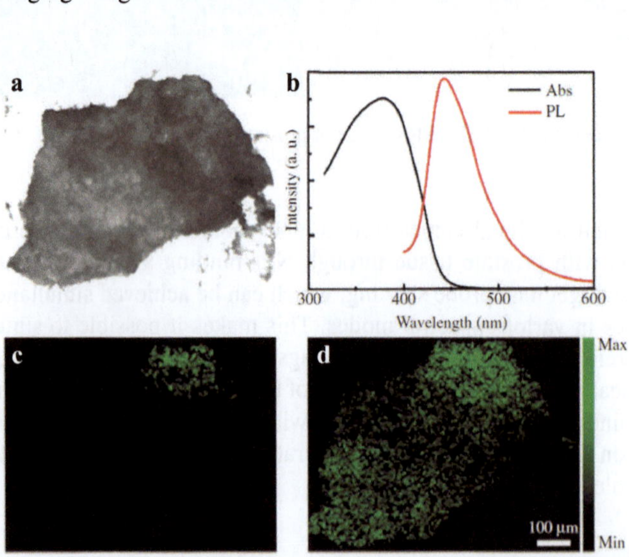

Fig. 8.4 Nonlinear optics imaging of $g - C_3N_4$. **A** Bright-field image of $g - C_3N_4$, **B** absorption and PL spectra of $g - C_3N_4$, **C** TPEF of $g - C_3N_4$, and **D** plasmon-enhanced TPEF of $g - C_3N_4$ with SPR [82]

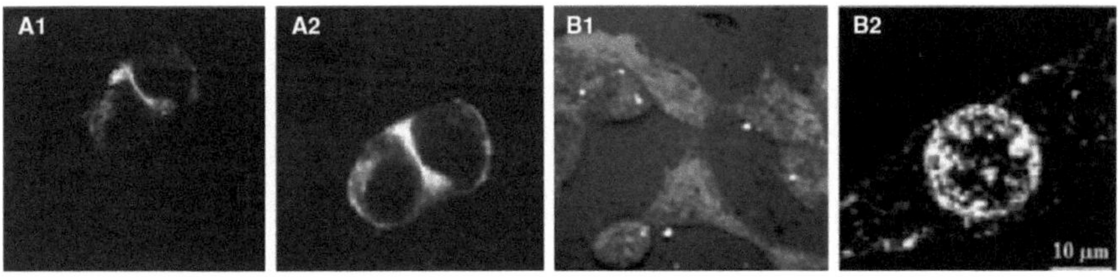

Fig. 8.5 Surface plasmon enhanced SHG and THG image for neuroblastoma cells. (A1, A2) With and Without SPR enhanced SHG signal by Au nanoparticles. (B1, B2) With and Without SPR enhanced THG signal by Au nanoparticles

in the cytoplasmic area in the SHG image after the surface plasmon enhancement, while the THG signal is strong in the nuclear area. This is due to the large third-order nonlinear coefficients in the highly ordered helical structure of DNA.

References

1. M. Rajadhyaksha, S. González, J.M. Zavislan, R.R. Anderson, R.H. Webb, In vivo confocal scanning laser microscopy of human skin ii: advances in instrumentation and comparison with histology. J. Investig. Dermatol. **113**(3), 293–303 (1999)
2. J.G. Fujimoto, Optical coherence tomography for ultrahigh resolution in vivo imaging. Nat. Biotechnol. **21**(11), 1361 (2003)
3. W.R. Zipfel, R.M. Williams, R. Christie, A.Y. Nikitin, B.T. Hyman, W.W. Webb, Live tissue intrinsic emission microscopy using multiphoton-excited native fluorescence and second harmonic generation. Proc. Natl. Acad. Sci. **100**(12), 7075–7080 (2003)
4. P.J. Campagnola, L.M. Loew, Second-harmonic imaging microscopy for visualizing biomolecular arrays in cells, tissues and organisms. Nat. Biotechnol. **21**(11), 1356 (2003)
5. L.M. Miller, G.D. Smith, G.L. Carr, Synchrotron-based biological microspectroscopy: from the mid-infrared through the far-infrared regimes. J. Biol. Phys. **29**(2–3), 219–230 (2003)
6. P.D. Maker, R.W. Terhune, Study of optical effects due to an induced polarization third order in the electric field strength. Phys. Rev. **137**(3A), A801 (1965)
7. W. Denk, J.H. Strickler, W.W. Webb, Two-photon laser scanning fluorescence microscopy. Science **248**(4951), 73–76 (1990)
8. J.-X. Cheng, X.S. Xie, Coherent anti-stokes Raman scattering microscopy: instrumentation, theory, and applications (2004)
9. X. Nan, J.-X. Cheng, X.S. Xie, Vibrational imaging of lipid droplets in live fibroblast cells with coherent anti-stokes Raman scattering microscopy. J. Lipid Res. **44**(11), 2202–2208 (2003)
10. J.-X. Cheng, L.D. Book, X.S. Xie, Polarization coherent anti-stokes Raman scattering microscopy. Opt. Lett. **26**(17), 1341–1343 (2001)
11. E.O. Potma, W.P. de Boeij, P.J.M. van Haastert, D.A. Wiersma, Real-time visualization of intracellular hydrodynamics in single living cells. Proc. Natl. Acad. Sci. **98**(4), 1577–1582 (2001)
12. M. Rajadhyaksha, M. Grossman, D. Esterowitz, R.H. Webb, R.R. Anderson, In vivo confocal scanning laser microscopy of human skin: melanin provides strong contrast. J. Investig. Dermatol. (1995)
13. E. Betzig, G.H. Patterson, R.S.O.W. Lindwasser, S. Olenych, J.S. Bonifacino, M.W. Davidson, J. Lippincott-Schwartz, H.F. Hess, Imaging intracellular fluorescent proteins at nanometer resolution. Science **313**(5793), 1642–1645 (2006)

M. Sun et al., *Linear and Nonlinear Optical Spectroscopy and Microscopy*,
Progress in Optical Science and Photonics 29,
https://doi.org/10.1007/978-981-99-3637-3

14. M.J. Rust, M. Bates, X. Zhuang, Stochastic optical reconstruction microscopy (STORM) provides sub-diffraction-limit image resolution. Nat. Methods **3**(10), 793 (2006)
15. W.P. Dempsey, S.E. Fraser, P. Pantazis, SHG nanoprobes: advancing harmonic imaging in biology. Bioessays News Rev. Mol. Cell. Dev. Biol. **34**(5), 351–360 (2012)
16. D. Staedler, T. Magouroux, R. Hadji, C. Joulaud, J.P. Wolf, Harmonic nanocrystals for biolabeling: a survey of optical properties and biocompatibility. ACS Nano **6**(3), 2542–2549 (2012)
17. C.-L. Sun, J.L.X.-Z. Wang, R. Shen, S. Liu, J.-Q. Jiang, T. Li, Q.-W. Song, Q. Liao, H.-B. Fu, Rational design of organic probes for turn-on two-photon excited fluorescence imaging and photodynamic therapy. Chem (2019)
18. P. Periklis, J. Maloney, D. Wu, S.E. Fraser, Second harmonic generating (SHG) nanoprobes for in vivo imaging. Proc. Natl. Acad. Sci. U. S. A. (2010)
19. J.G. Croissant, J.I. Zink, L. Raehm, J.-O. Durand, Two-photon-excited silica and organosilica nanoparticles for spatiotemporal cancer treatment. Adv. Healthc. Mater. **7**(7), 1701248 (2018)
20. C.L. Evans, X.S. Xie, Coherent anti-stokes raman scattering microscopy: chemical imaging for biology and medicine. Annu. Rev. Anal. Chem. **1**, 883–909 (2008)
21. A. Zumbusch, G.R. Holtom, X.S. Xie, Three-dimensional vibrational imaging by coherent anti-stokes Raman scattering. Phys. Rev. Lett. **82**(20), 4142 (1999)
22. W. Stummer, J.-C. Tonn, H.M. Mehdorn, U. Nestler, K. Franz, C. Goetz, A. Bink, U. Pichlmeier, Counterbalancing risks and gains from extended resections in malignant glioma surgery: a supplemental analysis from the randomized 5-aminolevulinic acid glioma resection study. J. Neurosurg. **114**(3), 613–623 (2011)
23. W. Chen, Clinical applications of pet in brain tumors. J. Nucl. Med. **48**(9), 1468–1481 (2007)
24. M. Makary, E.A. Chiocca, N. Erminy, M. Antor, S.D. Bergese, M. Abdel-Rasoul, S. Fernandez, R. Dzwonczyk, Clinical and economic outcomes of low-field intraoperative MRI-guided tumor resection neurosurgery. J. Magn. Reson. Imaging **34**(5), 1022–1030 (2011)
25. J. Regelsberger, F. Lohmann, K. Helmke, M. Westphal, Ultrasound-guided surgery of deep seated brain lesions. Eur. J. Ultrasound **12**(2), 115–121 (2000)
26. H.J. Böhringer, E. Lankenau, F. Stellmacher, E. Reusche, G. Hüttmann, A. Giese, Imaging of human brain tumor tissue by near-infrared laser coherence tomography. Acta Neurochir. **151**(5), 507–517 (2009)
27. N. Sanai, J. Eschbacher, G. Hattendorf, S.W. Coons, M.C. Preul, K.A. Smith, P. Nakaji, R.F. Spetzler, Intraoperative confocal microscopy for brain tumors: a feasibility analysis in humans. Oper. Neurosurg. **68**(suppl_2):ons282–ons290 (2011)
28. W. Stummer, J.-C. Tonn, C. Goetz, W. Ullrich, H. Stepp, A. Bink, T. Pietsch, U. Pichlmeier, 5-aminolevulinic acid-derived tumor fluorescence: the diagnostic accuracy of visible fluorescence qualities as corroborated by spectrometry and histology and postoperative imaging. Neurosurgery **74**(3), 310–320 (2014)
29. P. Kremer, M. Fardanesh, R. Ding, M. Pritsch, S. Zoubaa, E. Frei, Intraoperative fluorescence staining of malignant brain tumors using 5-aminofluorescein-labeled albumin. Oper. Neurosurg. **64**(suppl_1):ONS53–ONS61 (2009)
30. D.A. Dombeck, K.A. Kasischke, H.D. Vishwasrao, M. Ingelsson, B.T. Hyman, W.W. Webb, Uniform polarity microtubule assemblies imaged in native brain tissue by second-harmonic generation microscopy. Proc. Natl. Acad. Sci. **100**(12), 7081–7086 (2003)
31. S. Witte, A. Negrean, J.C. Lodder, C.P.J. De Kock, G.T. Silva, H.D. Mansvelder, M.L. Groot, Label-free live brain imaging and targeted patching with third-harmonic generation microscopy. Proc. Natl. Acad. Sci. **108**(15), 5970–5975 (2011)
32. Z. Movasaghi, S. Rehman, I.U. Rehman, Raman spectroscopy of biological tissues. Appl. Spectrosc. Rev. **42**(5), 493–541 (2007)
33. R. Bhargava, Infrared spectroscopic imaging: the next generation. Appl. Spectrosc. **66**(10), 1091–1120 (2012)
34. T. Meyer, N. Bergner, C. Krafft, D. Akimov, B. Dietzek, J. Popp, C. Bielecki, B.F.M. Romeike, R. Reichart, R. Kalff, Nonlinear microscopy, infrared, and Raman microspectroscopy for brain tumor analysis. J. Biomed. Opt. **16**(2), 021113 (2011)

35. H. Karabeber, R. Huang, P. Iacono, J.M. Samii, K. Pitter, E.C. Holland, M.F. Kircher, Guiding brain tumor resection using surface-enhanced Raman scattering nanoparticles and a hand-held Raman scanner. ACS Nano **8**(10), 9755–9766 (2014)
36. F.-K. Lu, S. Basu, V. Igras, M.P. Hoang, M. Ji, D. Fu, G.R. Holtom, V.A. Neel, C.W. Freudiger, D.E. Fisher, et al., Label-free DNA imaging in vivo with stimulated Raman scattering microscopy. Proc. Natl. Acad. Sci. **112**(37), 11624–11629 (2015)
37. J. Seidel, Y. Miao, W. Porterfield, W. Cai, X. Zhu, S.-J. Kim, H. Fanghao, S. Bhattarai-Kline, W. Min, W. Zhang, Structure-activity-distribution relationship study of anti-cancer antimycin-type depsipeptides. Chem. Commun. **55**(63), 9379–9382 (2019)
38. C.-L. Hsieh, R. Grange, Y. Pu, D. Psaltis, Three-dimensional harmonic holographic microcopy using nanoparticles as probes for cell imaging. Opt. Express **17**(4), 2880–2891 (2009)
39. S. Vanden-Hehir, S.A. Cairns, M. Lee, L. Zoupi, M.P. Shaver, V.G. Brunton, A. Williams, A.N. Hulme, Alkyne-tagged PLGA allows direct visualization of nanoparticles in vitro and ex vivo by stimulated Raman scattering microscopy. Biomacromolecules **20**(10), 4008–4014 (2019)
40. B. Huang, S. Yan, L. Xiao, R. Ji, L. Yang, A.-J. Miao, P. Wang, Label-free imaging of nanoparticle uptake competition in single cells by hyperspectral stimulated Raman scattering. Small **14**(10), 1703246 (2018)
41. F.-K. Lu, D. Calligaris, O.I. Olubiyi, I. Norton, W. Yang, S. Santagata, X.S. Xie, A.J. Golby, N.Y.R. Agar, Label-free neurosurgical pathology with stimulated Raman imaging. Cancer Res. **76**(12), 3451–3462 (2016)
42. O. Uckermann, R. Galli, S. Tamosaityte, E. Leipnitz, K.D. Geiger, G. Schackert, E. Koch, G. Steiner, M. Kirsch, Label-free delineation of brain tumors by coherent anti-stokes Raman scattering microscopy in an orthotopic mouse model and human glioblastoma. PLoS One **9**(9), e107115 (2014)
43. D.A. Orringer, B. Pandian, Y.S. Niknafs, T.C. Hollon, J. Boyle, S. Lewis, M. Garrard, S.L. Hervey-Jumper, H.J.L. Garton, C.O. Maher et al., Rapid intraoperative histology of unprocessed surgical specimens via fibre-laser-based stimulated Raman scattering microscopy. Nat. Biomed. Eng. **1**(2), 0027 (2017)
44. D. Fu, F.-K. Lu, X. Zhang, C. Freudiger, D.R. Pernik, G. Holtom, X.S. Xie, Quantitative chemical imaging with multiplex stimulated Raman scattering microscopy. J. Am. Chem. Soc. **134**(8), 3623–3626 (2012)
45. J.-X. Cheng, X.S. Xie, Vibrational spectroscopic imaging of living systems: an emerging platform for biology and medicine. Science **350**(6264) (2015)
46. F. Helmchen, W. Denk, Deep tissue two-photon microscopy. Nat. Methods **2**(12), 932–940 (2005)
47. M.J. Pittet, R. Weissleder, Intravital imaging. Cell **147**(5), 983–991 (2011)
48. B.-G. Wang, K. König, K.-J. Halbhuber, Two-photon microscopy of deep intravital tissues and its merits in clinical research. J. Microsc. **238**(1), 1–20 (2010)
49. D. Kantere, S. Guldbrand, J. Paoli, M. Goksör, D. Hanstorp, A.-M. Wennberg, M. Smedh, M.B. Ericson, Anti-stokes fluorescence from endogenously formed protoporphyrin ix-implications for clinical multiphoton diagnostics. J. Biophotonics **6**(5), 409–415 (2013)
50. M.Y. Berezin, S. Achilefu, Fluorescence lifetime measurements and biological imaging. Chem. Rev. **110**(5), 2641–2684 (2010)
51. L. Marcu, Fluorescence lifetime techniques in medical applications. Ann. Biomed. Eng. **40**(2), 304–331 (2012)
52. D. Chorvat Jr., A. Chorvatova, Multi-wavelength fluorescence lifetime spectroscopy: a new approach to the study of endogenous fluorescence in living cells and tissues. Laser Phys. Lett. **6**(3), 175–193 (2009)
53. B.R. Masters, P. So, *Handbook of Biomedical Nonlinear Optical Microscopy* (Oxford University Press, 2008)
54. K. König, Clinical multiphoton tomography. J. Biophotonics **1**(1), 13–23 (2008)
55. J.R. Lakowicz, *Principles of Fluorescence Spectroscopy* (Springer Science & Business Media, 2013)

56. L.K. Seah, P. Wang, V.M. Murukeshan, Z.X. Chao, Application of fluorescence lifetime imaging (FLIM) in latent finger mark detection. Forensic Sci. Int. **160**(2–3), 109–114 (2006)
57. A. Pliss, X. Peng, L. Liu, A. Kuzmin, Y. Wang, J. Qu, Y. Li, P. Prasad, Single cell assay for molecular diagnostics and medicine: monitoring intracellular concentrations of macromolecules by two-photon fluorescence lifetime imaging. Theranostics **5**(9), 919–930 (2015)
58. M.A. Baldo, D.F. O'brien, Y. You, A. Shoustikov, S. Sibley, M.E. Thompson, S.R. Forrest, Highly efficient phosphorescent emission from organic electroluminescent devices. Nature **395**(6698), 151 (1998)
59. M.J. Frisch, G.W. Trucks, H.B. Schlegel, G.E. Scuseria, M.A. Robb, J.R. Cheeseman, G. Scalmani, V. Barone, G.A. Petersson, H.J.R.A. Nakatsuji, et al., Gaussian 16 (2016)
60. L. Tian, F. Chen, Multiwfn: a multifunctional wavefunction analyzer. J. Comput. Chem. **33**(5), 580–592 (2012)
61. K. Sasagane, F. Aiga, R. Itoh, Higher-order response theory based on the quasienergy derivatives: the derivation of the frequency-dependent polarizabilities and hyperpolarizabilities. J. Chem. Phys. **99**(5), 3738–3778 (1993)
62. S. Hirata, M. Head-Gordon, Time-dependent density functional theory within the Tamm-Dancoff approximation. Chem. Phys. Lett. **314**(3–4), 291–299 (1999)
63. R.W. Boyd, *Nonlinear Optics* (Elsevier, 2003)
64. T.F. Heinz, C.K. Chen, D. Ricard, Y.R. Shen, Spectroscopy of molecular monolayers by resonant second-harmonic generation. Phys. Rev. Lett. **48**(7), 478 (1982)
65. M. Han, G. Giese, J.F. Bille, Second harmonic generation imaging of collagen fibrils in cornea and sclera. Opt. Express **13**(15), 5791–5797 (2005)
66. M. Nuriya, S. Fukushima, A. Momotake, T. Shinotsuka, M. Yasui, T. Arai, Multimodal two-photon imaging using a second harmonic generation-specific dye. Nat. Commun. **7**, 11557 (2016)
67. A. Khadria, Y. de Coene, P. Gawel, C. Roche, K. Clays, H.L. Anderson, Push-pull pyropheophorbides for nonlinear optical imaging. Org. Biomol. Chem. **15**(4), 947–956 (2017)
68. V.K. Valev, Characterization of nanostructured plasmonic surfaces with second harmonic generation. Langmuir **28**(44), 15454–15471 (2012)
69. W. M_ Tolles, J.W. Nibler, J.R. McDonald, A.B. Harvey, A review of the theory and application of coherent anti-stokes Raman spectroscopy (cars). Appl. Spectrosc. **31**(4), 253–271 (1977)
70. R. Li, L. Wang, M. Xijiao, M. Chen, M. Sun, Biological nascent evolution of snail bone and collagen revealed by nonlinear optical microscopy. J. Biophotonics **12**(8), e201900119 (2019)
71. R. Li, Y. Zhang, X. Xuefeng, Y. Zhou, M. Chen, M. Sun, Optical characterizations of two-dimensional materials using nonlinear optical microscopies of CARS, TPEF, and SHG. Nanophotonics **7**(5), 873–881 (2018)
72. C.W. Freudiger, W. Min, B.G. Saar, S. Lu, G.R. Holtom, C. He, J.C. Tsai, J.X. Kang, X.S. Xie, Label-free biomedical imaging with high sensitivity by stimulated Raman scattering microscopy. Science **322**(5909), 1857–1861 (2008)
73. A. Folick, W. Min, M.C. Wang, Label-free imaging of lipid dynamics using coherent anti-stokes Raman scattering (CARS) and stimulated Raman scattering (SRS) microscopy. Curr. Opin. Genet. Dev. **21**(5), 585–590 (2011)
74. C.L. Evans, X. Xu, S. Kesari, X.S. Xie, S.T.C. Wong, G.S. Young, Chemically-selective imaging of brain structures with cars microscopy. Opt. Express **15**(19), 12076–12087 (2007)
75. M. Ji, S. Lewis, S. Camelo-Piragua, S.H. Ramkissoon, M. Snuderl, S. Venneti, A. Fisher-Hubbard, M. Garrard, D. Fu, A.C. Wang, et al., Detection of human brain tumor infiltration with quantitative stimulated Raman scattering microscopy. Sci. Transl. Med. **7**(309), 309ra163–309ra163 (2015)
76. L. Zhang, W. Yongzheng, B. Zheng, S. Lizhong, Y. Chen, S. Ma, H. Qinqin, X. Zou, L. Yao, Y. Yang, et al., Rapid histology of laryngeal squamous cell carcinoma with deep-learning based stimulated Raman scattering microscopy. Theranostics **9**(9), 2541 (2019)
77. A. Blázquez-Castro, J.C. Stockert, *Fluorescence Microscopy in Life Sciences* (Bentham Science Publishers, 2017)

78. V. Nucciotti, C. Stringari, L. Sacconi, F. Vanzi, L. Fusi, M. Linari, G. Piazzesi, V. Lombardi, F.S. Pavone, Probing myosin structural conformation in vivo by second-harmonic generation microscopy. Proc. Natl. Acad. Sci. (2010)
79. P.J. Campagnola, X. Chen, SHG microscopy and its comparison with THG, CARS, and multi-photon excited fluorescence imaging. In: *Second Harmonic Generation Imaging* (CRC Press, 2016), pp. 100–119
80. L. Mostaço-Guidolin, N.L. Rosin, T.-L. Hackett, Imaging collagen in scar tissue: developments in second harmonic generation microscopy for biomedical applications. Int. J. Mol. Sci. **18**(8), 1772 (2017)
81. X. Chen, O. Nadiarynkh, S. Plotnikov, P.J. Campagnola, Second harmonic generation microscopy for quantitative analysis of collagen fibrillar structure. Nat. Protoc. **7**(4), 654–669 (2012)
82. X. Mi, Y. Wang, R. Li, M. Sun, Z. Zhang, H. Zheng, Multiple surface plasmon resonances enhanced nonlinear optical microscopy. Nanophotonics **8**(3), 487–493 (2019)
83. L. Sun, W. Zhu, W. Wang, F. Yang, C. Zhang, S. Wang, X. Zhang, R. Li, H. Dong, H. Wenping, Intermolecular charge-transfer interactions facilitate two-photon absorption in styrylpyridine-tetracyanobenzene cocrystals. Angew. Chem. Int. Ed. **56**(27), 7831–7835 (2017)
84. K.E. Dorfman, F. Schlawin, S. Mukamel, Nonlinear optical signals and spectroscopy with quantum light. Rev. Mod. Phys. **88**(4), 045008 (2016)
85. J. Herz, V. Siffrin, A.E. Hauser, A.U. Brandt, T. Leuenberger, H. Radbruch, F. Zipp, R.A. Niesner, Expanding two-photon intravital microscopy to the infrared by means of optical parametric oscillator. Biophys. J. **98**(4), 715–723 (2010)
86. D. Entenberg, J. Wyckoff, B. Gligorijevic, E.T. Roussos, V.V. Verkhusha, J.W. Pollard, J. Condeelis, Setup and use of a two-laser multiphoton microscope for multichannel intravital fluorescence imaging. Nat. Protoc. **6**(10), 1500–1520 (2011)
87. C. Li, R.K. Pastila, C.P. Lin, Label-free imaging immune cells and collagen in atherosclerosis with two-photon and second harmonic generation microscopy. J. Innov. Opt. Health Sci. **09**(01), 1640003 (2016)
88. M.H. Brenner, D. Cai, J.A. Swanson, J.P. Ogilvie, Two-photon imaging of multiple fluorescent proteins by phase-shaping and linear unmixing with a single broadband laser. Opt. Express **21**(14), 17256–17264 (2013)
89. R.S. Pillai, C. Boudoux, G. Labroille, N. Olivier, I. Veilleux, E. Farge, M. Joffre, E. Beaurepaire, Multiplexed two-photon microscopy of dynamic biological samples with shaped broadband pulses. Opt. Express **17**(15), 12741–12752 (2009)
90. S.K. Teh, W. Zheng, S. Li, D. Li, Y. Zeng, Y. Yang, J.Y. Qu, Multimodal nonlinear optical microscopy improves the accuracy of early diagnosis of squamous intraepithelial neoplasia. J. Biomed. Opt. **18**(3), 036001 (2013)
91. B.R. Masters, P. So, *Handbook of Biomedical Nonlinear Optical Microscopy* (Oxford University Press, 2008)
92. W. Becker, *Advanced Time-Correlated Single Photon Counting Techniques*, vol. 81 (Springer Science & Business Media, 2005)
93. Y. Wu, W. Zheng, J.Y. Qu, Sensing cell metabolism by time-resolved autofluorescence. Opt. Lett. **31**(21), 3122–4 (2006)
94. U.V. Nägerl, K.I. Willig, B. Hein, S.W. Hell, Live-cell imaging of dendritic spines by STED microscopy. Proc. Natl. Acad. Sci. U. S. A. (2008)
95. J.F. Li, Y.F. Huang, Y. Ding, Z.L. Yang, S.B. Li, X.S. Zhou, F.R. Fan, W. Zhang, Z.Y. Zhou, B. Ren, et al., Shell-isolated nanoparticle-enhanced Raman spectroscopy. Nature **464**(7287), 392–395 (2010)
96. A. Otto, The 'chemical' (electronic) contribution to surface-enhanced Raman scattering. J. Raman Spectrosc.: Int. J. Orig. Work. All Asp. Raman Spectrosc., Incl. High. Order Process., Also Brillouin Rayleigh Scatt. **36**(6–7), 497–509 (2005)
97. D. Yelin, D. Oron, S. Thiberge, E. Moses, Y. Silberberg, Multiphoton plasmon-resonance microscopy. Opt. Express **11**(12), 1385–1391 (2003)

MIX
Papier aus verantwortungsvollen Quellen
Paper from responsible sources
FSC® C105338

If you have any concerns about our products,
you can contact us on
ProductSafety@springernature.com

In case Publisher is established outside the EU,
the EU authorized representative is:
Springer Nature Customer Service Center GmbH
Europaplatz 3, 69115 Heidelberg, Germany

Printed by Libri Plureos GmbH
in Hamburg, Germany